This Book Belongs To:

. . . .

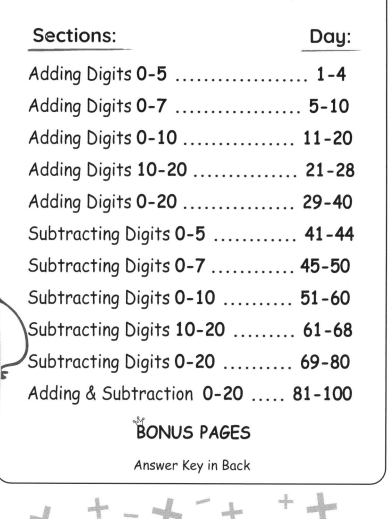

Table of Content

BONUS PAGES

Answer Key in Back

This math practice workbook is organized in a progressively skill building way for kids to develop confidence in Addition & Subtraction

ISBN: 9798683600716

1.	2.	3.	4.	5.	6.	7.	8.
5 + 2	5 + 4	2 + 1	1 + 2	4 + 1	0 + 2	2 + 0	2 + 1

9.	10.	11.	12.	13.	14.	15.	16.
2 + 1	5 + 3	1 + 0	5 + 3	1 + 0	3 + 4	3 + 3	2 + 2

17.	18.	19.	20.	21.	22.	23.	24.
2 + 2	1 + 3	2 + 4	1 + 5	4 + 4	3 + 2	5 + 5	0 + 3

25.	26.	27.	28.	29.	30.	31.	32.
4 + 0	0 + 4	4 + 3	0 + 3	0 + 5	4 + 1	4 + 2	4 + 0

33.	34.	35.	36.	37.	38.	39.	40.
4 + 0	2 + 1	3 + 5	3 + 2	1 + 4	0 + 0	2 + 4	4 + 5

41.	42.	43.	44.	45.	46.	47.	48.
5 + 4	3 + 3	1 + 4	3 + 3	2 + 5	3 + 0	1 + 5	4 + 4

49.	50.	51.	52.	53.	54.	55.	56.
5 + 0	1 + 1	5 + 1	3 + 4	0 + 5	3 + 3	3 + 3	5 + 1

57.	58.	59.	60.
1 + 4	0 + 1	2 + 3	3 + 1

Score /60   I Feel: ☹ 😐 ☺

1. 2
   + 2
   ___

2. 0
   + 1
   ___

3. 0
   + 3
   ___

4. 4
   + 5
   ___

5. 5
   + 2
   ___

6. 1
   + 1
   ___

7. 3
   + 2
   ___

8. 2
   + 5
   ___

9. 3
   + 2
   ___

10. 4
    + 1
    ___

11. 4
    + 0
    ___

12. 1
    + 4
    ___

13. 2
    + 3
    ___

14. 2
    + 0
    ___

15. 3
    + 5
    ___

16. 3
    + 4
    ___

17. 5
    + 0
    ___

18. 5
    + 4
    ___

19. 1
    + 2
    ___

20. 1
    + 3
    ___

21. 4
    + 3
    ___

22. 4
    + 2
    ___

23. 3
    + 0
    ___

24. 0
    + 2
    ___

25. 5
    + 4
    ___

26. 1
    + 0
    ___

27. 1
    + 2
    ___

28. 3
    + 0
    ___

29. 4
    + 2
    ___

30. 4
    + 2
    ___

31. 3
    + 1
    ___

32. 3
    + 4
    ___

33. 3
    + 1
    ___

34. 2
    + 2
    ___

35. 4
    + 3
    ___

36. 4
    + 4
    ___

37. 2
    + 3
    ___

38. 5
    + 0
    ___

39. 2
    + 3
    ___

40. 3
    + 4
    ___

41. 1
    + 5
    ___

42. 1
    + 3
    ___

43. 0
    + 0
    ___

44. 3
    + 3
    ___

45. 5
    + 1
    ___

46. 2
    + 4
    ___

47. 2
    + 1
    ___

48. 0
    + 5
    ___

49. 3
    + 3
    ___

50. 0
    + 4
    ___

51. 2
    + 3
    ___

52. 5
    + 5
    ___

53. 1
    + 2
    ___

54. 3
    + 1
    ___

55. 1
    + 2
    ___

56. 2
    + 4
    ___

57. 5
    + 2
    ___

58. 4
    + 4
    ___

59. 5
    + 5
    ___

60. 5
    + 3
    ___

Score /60

I Feel: ☹ 😐 🙂

Adding Digits 0 - 5

 Start   Finish

Date: ....   NAME : ....   Time: .....:..... .....:.....

1. 2 + 1	2. 5 + 5	3. 5 + 1	4. 3 + 4	5. 4 + 3	6. 3 + 4	7. 3 + 2	8. 5 + 2
9. 2 + 4	10. 4 + 2	11. 0 + 3	12. 3 + 1	13. 5 + 0	14. 5 + 4	15. 0 + 4	16. 4 + 4
17. 2 + 2	18. 1 + 3	19. 1 + 2	20. 2 + 3	21. 2 + 3	22. 1 + 3	23. 4 + 1	24. 0 + 2
25. 1 + 5	26. 4 + 2	27. 1 + 1	28. 4 + 3	29. 2 + 4	30. 4 + 0	31. 1 + 4	32. 1 + 2
33. 3 + 4	34. 5 + 0	35. 3 + 3	36. 4 + 2	37. 5 + 5	38. 2 + 3	39. 1 + 0	40. 3 + 3
41. 3 + 0	42. 5 + 3	43. 3 + 5	44. 0 + 0	45. 1 + 2	46. 4 + 4	47. 3 + 2	48. 1 + 2
49. 2 + 5	50. 5 + 2	51. 2 + 0	52. 3 + 1	53. 0 + 1	54. 3 + 0	55. 0 + 5	56. 4 + 5
57. 3 + 1	58. 2 + 2	59. 2 + 3	60. 5 + 4				

Score /60   I Feel: 😞 😐 🙂

Adding Digits 0 - 5

 Start   Finish

Date: ....   NAME : ....   Time: .....:..... .....:.....

| 1. 4
 + 2 | 2. 4
 + 4 | 3. 5
 + 4 | 4. 3
 + 3 | 5. 0
 + 3 | 6. 4
 + 2 | 7. 2
 + 1 | 8. 3
 + 3 |

| 9. 5
 + 3 | 10. 0
 + 4 | 11. 5
 + 2 | 12. 1
 + 3 | 13. 5
 + 0 | 14. 3
 + 4 | 15. 4
 + 1 | 16. 1
 + 2 |

| 17. 3
 + 1 | 18. 4
 + 3 | 19. 0
 + 5 | 20. 5
 + 1 | 21. 1
 + 2 | 22. 2
 + 2 | 23. 5
 + 5 | 24. 3
 + 5 |

| 25. 2
 + 0 | 26. 2
 + 3 | 27. 2
 + 3 | 28. 2
 + 4 | 29. 2
 + 2 | 30. 4
 + 2 | 31. 5
 + 2 | 32. 3
 + 4 |

| 33. 0
 + 2 | 34. 1
 + 2 | 35. 0
 + 0 | 36. 3
 + 1 | 37. 2
 + 3 | 38. 2
 + 5 | 39. 5
 + 0 | 40. 3
 + 0 |

| 41. 5
 + 5 | 42. 1
 + 0 | 43. 5
 + 4 | 44. 3
 + 1 | 45. 3
 + 2 | 46. 3
 + 4 | 47. 1
 + 1 | 48. 4
 + 4 |

| 49. 2
 + 3 | 50. 4
 + 3 | 51. 1
 + 4 | 52. 4
 + 0 | 53. 3
 + 0 | 54. 4
 + 5 | 55. 1
 + 2 | 56. 0
 + 1 |

| 57. 2
 + 4 | 58. 1
 + 5 | 59. 3
 + 2 | 60. 1
 + 3 |

Score /60

I Feel:

Day 5

Date: ....

NAME : ....

Start   Finish

Time: ....:....   ....:....

1.	2.	3.	4.	5.	6.	7.	8.
1 + 6	7 + 3	1 + 4	1 + 1	6 + 6	2 + 7	4 + 7	3 + 5

9.	10.	11.	12.	13.	14.	15.	16.
2 + 6	3 + 4	5 + 1	0 + 1	3 + 2	4 + 2	2 + 4	7 + 5

17.	18.	19.	20.	21.	22.	23.	24.
6 + 5	3 + 0	5 + 5	0 + 3	6 + 1	6 + 7	5 + 2	4 + 3

25.	26.	27.	28.	29.	30.	31.	32.
4 + 1	0 + 2	7 + 7	7 + 1	5 + 6	2 + 2	7 + 0	5 + 3

33.	34.	35.	36.	37.	38.	39.	40.
3 + 6	0 + 6	4 + 6	0 + 7	6 + 4	1 + 3	5 + 4	1 + 5

41.	42.	43.	44.	45.	46.	47.	48.
4 + 0	3 + 1	2 + 5	2 + 0	0 + 4	7 + 2	0 + 0	3 + 3

49.	50.	51.	52.	53.	54.	55.	56.
1 + 0	4 + 5	0 + 5	1 + 7	7 + 6	4 + 4	2 + 3	6 + 2

57.	58.	59.	60.
7 + 4	2 + 1	6 + 3	1 + 2

Score
/60

I Feel: 🙂

Adding Digits 0 - 7

Date: ....   NAME : ....

Start   Finish

Time: ....:....   ....:....

1. 4 + 7	2. 0 + 2	3. 3 + 1	4. 2 + 5	5. 3 + 6	6. 2 + 7	7. 7 + 3	8. 1 + 2
9. 6 + 4	10. 2 + 3	11. 1 + 0	12. 4 + 1	13. 2 + 6	14. 6 + 5	15. 3 + 4	16. 7 + 1
17. 1 + 6	18. 6 + 2	19. 6 + 1	20. 1 + 5	21. 3 + 0	22. 2 + 4	23. 0 + 5	24. 5 + 2
25. 0 + 0	26. 0 + 7	27. 3 + 2	28. 0 + 3	29. 4 + 5	30. 7 + 2	31. 4 + 0	32. 7 + 4
33. 3 + 5	34. 4 + 3	35. 2 + 2	36. 7 + 7	37. 3 + 3	38. 2 + 0	39. 6 + 6	40. 5 + 5
41. 7 + 0	42. 2 + 1	43. 1 + 7	44. 6 + 7	45. 5 + 6	46. 5 + 3	47. 4 + 6	48. 5 + 1
49. 1 + 1	50. 5 + 4	51. 0 + 4	52. 0 + 1	53. 1 + 3	54. 1 + 4	55. 7 + 6	56. 6 + 3
57. 0 + 6	58. 7 + 5	59. 4 + 4	60. 4 + 2				

Score /60

I Feel:

Adding Digits 0 - 7

Start   Finish

Date: ....   NAME : ....   Time: .....:..... ....:.....

1.   0
  + 4

2.   1
  + 6

3.   5
  + 2

4.   0
  + 2

5.   3
  + 6

6.   3
  + 3

7.   4
  + 1

8.   6
  + 2

9.   1
  + 4

10.   1
  + 0

11.   1
  + 2

12.   7
  + 5

13.   4
  + 4

14.   2
  + 3

15.   2
  + 0

16.   7
  + 1

17.   3
  + 1

18.   3
  + 4

19.   6
  + 7

20.   6
  + 3

21.   7
  + 2

22.   2
  + 1

23.   4
  + 6

24.   0
  + 5

25.   2
  + 5

26.   5
  + 3

27.   4
  + 7

28.   0
  + 1

29.   1
  + 3

30.   5
  + 5

31.   2
  + 6

32.   0
  + 7

33.   2
  + 2

34.   4
  + 0

35.   0
  + 3

36.   4
  + 3

37.   0
  + 0

38.   3
  + 5

39.   4
  + 5

40.   1
  + 1

41.   6
  + 6

42.   7
  + 6

43.   7
  + 4

44.   1
  + 7

45.   0
  + 6

46.   6
  + 5

47.   6
  + 1

48.   7
  + 0

49.   5
  + 1

50.   3
  + 2

51.   1
  + 5

52.   3
  + 0

53.   2
  + 7

54.   4
  + 2

55.   6
  + 4

56.   5
  + 4

57.   7
  + 3

58.   2
  + 4

59.   5
  + 6

60.   7
  + 7

Score /60

I Feel:

Adding Digits 0 - 7

Date: ....   NAME : ....

Start   Finish

Time: ....:....   ....:....

1. 2 + 5	2. 6 + 5	3. 5 + 0	4. 6 + 4	5. 7 + 7	6. 4 + 2	7. 2 + 1	8. 1 + 1
9. 3 + 4	10. 1 + 3	11. 5 + 7	12. 3 + 4	13. 2 + 1	14. 6 + 4	15. 7 + 0	16. 3 + 1
17. 4 + 6	18. 6 + 0	19. 7 + 2	20. 5 + 3	21. 6 + 6	22. 2 + 0	23. 2 + 7	24. 7 + 6
25. 2 + 4	26. 3 + 6	27. 3 + 5	28. 7 + 1	29. 1 + 2	30. 7 + 3	31. 1 + 6	32. 3 + 0
33. 5 + 5	34. 4 + 5	35. 5 + 6	36. 2 + 2	37. 2 + 6	38. 7 + 4	39. 3 + 7	40. 1 + 7
41. 5 + 1	42. 6 + 1	43. 6 + 2	44. 4 + 0	45. 1 + 5	46. 3 + 2	47. 4 + 3	48. 5 + 2
49. 4 + 1	50. 2 + 3	51. 5 + 4	52. 6 + 7	53. 4 + 7	54. 3 + 3	55. 4 + 5	56. 6 + 3
57. 1 + 0	58. 4 + 4	59. 7 + 5	60. 1 + 4				

Score /60

I Feel: :)

Day 9

Adding Digits 0 - 7

Start   Finish

Date: ....   NAME : ....   Time: .....:..... .....:.....

1. 6
   + 5

2. 5
   + 7

3. 1
   + 6

4. 2
   + 4

5. 7
   + 3

6. 0
   + 1

7. 1
   + 7

8. 7
   + 2

9. 7
   + 7

10. 5
    + 2

11. 0
    + 5

12. 5
    + 6

13. 0
    + 3

14. 7
    + 4

15. 4
    + 1

16. 0
    + 4

17. 3
    + 3

18. 2
    + 6

19. 5
    + 5

20. 3
    + 1

21. 4
    + 3

22. 3
    + 5

23. 4
    + 4

24. 7
    + 5

25. 5
    + 4

26. 7
    + 1

27. 5
    + 1

28. 3
    + 6

29. 3
    + 2

30. 7
    + 5

31. 6
    + 4

32. 6
    + 2

33. 1
    + 2

34. 1
    + 1

35. 6
    + 1

36. 1
    + 3

37. 4
    + 6

38. 2
    + 1

39. 2
    + 2

40. 6
    + 3

41. 5
    + 2

42. 7
    + 1

43. 3
    + 4

44. 1
    + 4

45. 2
    + 7

46. 5
    + 3

47. 4
    + 2

48. 1
    + 5

49. 0
    + 7

50. 4
    + 5

51. 0
    + 6

52. 6
    + 6

53. 3
    + 7

54. 6
    + 7

55. 2
    + 5

56. 0
    + 2

57. 4
    + 7

58. 7
    + 4

59. 2
    + 3

60. 7
    + 6

Score
/60

I Feel: 😞 😐 🙂

Start   Finish

Day 10   Date: ....   NAME : ....   Time: ....:.... ....:....

1. 1
   + 6

2. 6
   + 5

3. 3
   + 3

4. 4
   + 1

5. 1
   + 1

6. 2
   + 6

7. 7
   + 6

8. 0
   + 2

9. 7
   + 1

10. 3
    + 2

11. 4
    + 3

12. 0
    + 1

13. 3
    + 0

14. 4
    + 5

15. 5
    + 1

16. 6
    + 3

17. 5
    + 6

18. 0
    + 7

19. 5
    + 5

20. 2
    + 0

21. 4
    + 4

22. 3
    + 4

23. 0
    + 4

24. 7
    + 5

25. 6
    + 0

26. 6
    + 7

27. 2
    + 1

28. 1
    + 0

29. 3
    + 5

30. 2
    + 5

31. 4
    + 2

32. 6
    + 4

33. 1
    + 4

34. 1
    + 2

35. 4
    + 7

36. 1
    + 3

37. 7
    + 4

38. 6
    + 6

39. 6
    + 1

40. 7
    + 7

41. 5
    + 2

42. 6
    + 2

43. 0
    + 3

44. 7
    + 3

45. 3
    + 7

46. 5
    + 4

47. 2
    + 4

48. 2
    + 3

49. 7
    + 2

50. 3
    + 6

51. 5
    + 3

52. 3
    + 1

53. 2
    + 2

54. 2
    + 7

55. 4
    + 0

56. 0
    + 6

57. 1
    + 5

58. 7
    + 0

59. 1
    + 7

60. 4
    + 6

Score
/60

I Feel: 😞 😐 🙂

1. 4
   + 2

2. 3
   + 5

3. 10
   + 3

4. 1
   + 4

5. 3
   + 1

6. 2
   + 2

7. 6
   + 1

8. 8
   + 5

9. 1
   + 3

10. 6
   + 5

11. 5
   + 8

12. 4
   + 8

13. 1
   + 7

14. 2
   + 1

15. 8
   + 8

16. 3
   + 9

17. 3
   + 8

18. 4
   + 1

19. 2
   + 4

20. 6
   + 7

21. 9
   + 4

22. 0
   + 5

23. 9
   + 2

24. 6
   + 8

25. 10
   + 2

26. 3
   + 7

27. 6
   + 4

28. 1
   + 5

29. 4
   + 10

30. 9
   + 0

31. 10
   + 4

32. 9
   + 6

33. 2
   + 7

34. 8
   + 6

35. 6
   + 9

36. 4
   + 3

37. 2
   + 6

38. 2
   + 8

39. 10
   + 10

40. 5
   + 7

41. 3
   + 10

42. 6
   + 0

43. 5
   + 9

44. 4
   + 6

45. 0
   + 7

46. 7
   + 1

47. 9
   + 8

48. 4
   + 9

49. 1
   + 1

50. 1
   + 6

51. 8
   + 1

52. 3
   + 4

53. 3
   + 3

54. 7
   + 6

55. 2
   + 5

56. 2
   + 3

57. 10
   + 9

58. 1
   + 9

59. 10
   + 6

60. 4
   + 4

Score
/60

I Feel:

1. 9
   + 6

2. 8
   + 6

3. 4
   + 9

4. 2
   + 7

5. 10
   + 10

6. 0
   + 7

7. 3
   + 9

8. 2
   + 1

9. 3
   + 7

10. 9
    + 4

11. 2
    + 5

12. 6
    + 1

13. 4
    + 8

14. 9
    + 8

15. 2
    + 8

16. 2
    + 6

17. 9
    + 2

18. 6
    + 9

19. 3
    + 4

20. 5
    + 8

21. 5
    + 9

22. 10
    + 6

23. 1
    + 3

24. 10
    + 2

25. 6
    + 5

26. 3
    + 8

27. 6
    + 0

28. 4
    + 2

29. 2
    + 4

30. 2
    + 2

31. 0
    + 5

32. 3
    + 1

33. 6
    + 7

34. 9
    + 0

35. 10
    + 3

36. 1
    + 6

37. 1
    + 1

38. 4
    + 6

39. 8
    + 5

40. 10
    + 9

41. 7
    + 6

42. 1
    + 9

43. 8
    + 1

44. 4
    + 3

45. 4
    + 4

46. 2
    + 3

47. 10
    + 4

48. 8
    + 8

49. 3
    + 3

50. 7
    + 1

51. 3
    + 5

52. 5
    + 7

53. 4
    + 10

54. 1
    + 4

55. 1
    + 5

56. 6
    + 8

57. 4
    + 1

58. 6
    + 4

59. 1
    + 7

60. 3
    + 10

Score /60

I Feel: ☹ 😐 🙂

1. 4
    + 3

2. 3
    + 1

3. 3
    + 7

4. 6
    + 6

5. 0
    + 5

6. 6
    + 9

7. 6
    + 8

8. 6
    + 2

9. 8
    + 1

10. 9
    + 1

11. 2
    + 7

12. 7
    + 10

13. 5
    + 10

14. 9
    + 8

15. 7
    + 5

16. 6
    + 1

17. 2
    + 9

18. 10
    + 6

19. 4
    + 1

20. 5
    + 9

21. 9
    + 6

22. 3
    + 8

23. 1
    + 8

24. 8
    + 3

25. 1
    + 5

26. 2
    + 6

27. 7
    + 1

28. 1
    + 6

29. 5
    + 8

30. 5
    + 2

31. 2
    + 10

32. 10
    + 5

33. 1
    + 2

34. 1
    + 9

35. 6
    + 4

36. 7
    + 9

37. 3
    + 4

38. 4
    + 8

39. 3
    + 10

40. 5
    + 7

41. 8
    + 6

42. 0
    + 7

43. 9
    + 4

44. 2
    + 2

45. 1
    + 7

46. 0
    + 3

47. 2
    + 1

48. 1
    + 3

49. 8
    + 5

50. 5
    + 5

51. 4
    + 5

52. 10
    + 0

53. 0
    + 8

54. 10
    + 3

55. 10
    + 1

56. 8
    + 7

57. 0
    + 2

58. 10
    + 8

59. 7
    + 8

60. 3
    + 5

Score /60

I Feel:

1. 6
+ 6

2. 1
+ 7

3. 2
+ 6

4. 5
+ 10

5. 8
+ 6

6. 7
+ 10

7. 4
+ 3

8. 0
+ 5

9. 10
+ 3

10. 6
+ 2

11. 2
+ 1

12. 1
+ 6

13. 1
+ 2

14. 2
+ 2

15. 8
+ 7

16. 0
+ 8

17. 5
+ 2

18. 8
+ 1

19. 1
+ 9

20. 3
+ 7

21. 4
+ 8

22. 3
+ 4

23. 7
+ 9

24. 2
+ 7

25. 4
+ 5

26. 10
+ 6

27. 9
+ 8

28. 1
+ 8

29. 1
+ 5

30. 10
+ 8

31. 0
+ 2

32. 2
+ 9

33. 3
+ 5

34. 4
+ 1

35. 7
+ 5

36. 2
+ 10

37. 5
+ 8

38. 5
+ 7

39. 6
+ 1

40. 8
+ 5

41. 9
+ 1

42. 0
+ 3

43. 5
+ 9

44. 10
+ 5

45. 3
+ 1

46. 7
+ 8

47. 6
+ 8

48. 6
+ 4

49. 9
+ 6

50. 6
+ 9

51. 5
+ 5

52. 8
+ 3

53. 0
+ 7

54. 3
+ 8

55. 1
+ 3

56. 9
+ 4

57. 7
+ 1

58. 3
+ 10

59. 10
+ 0

60. 10
+ 1

Score /60   I Feel:

1. 5
   + 10

2. 8
   + 5

3. 0
   + 3

4. 3
   + 10

5. 1
   + 6

6. 2
   + 9

7. 9
   + 8

8. 5
   + 2

9. 3
   + 5

10. 10
    + 6

11. 4
    + 5

12. 1
    + 3

13. 9
    + 4

14. 7
    + 8

15. 5
    + 5

16. 6
    + 7

17. 4
    + 8

18. 4
    + 3

19. 7
    + 10

20. 9
    + 6

21. 6
    + 2

22. 5
    + 9

23. 8
    + 7

24. 3
    + 7

25. 9
    + 5

26. 3
    + 1

27. 6
    + 4

28. 6
    + 9

29. 6
    + 1

30. 5
    + 7

31. 8
    + 1

32. 10
    + 3

33. 1
    + 2

34. 10
    + 5

35. 5
    + 8

36. 10
    + 8

37. 6
    + 6

38. 2
    + 8

39. 3
    + 4

40. 1
    + 5

41. 4
    + 1

42. 3
    + 8

43. 6
    + 2

44. 2
    + 10

45. 7
    + 2

46. 7
    + 1

47. 8
    + 6

48. 7
    + 5

49. 2
    + 2

50. 1
    + 8

51. 1
    + 7

52. 2
    + 6

53. 1
    + 9

54. 9
    + 1

55. 7
    + 9

56. 2
    + 7

57. 6
    + 8

58. 8
    + 3

59. 10
    + 1

60. 2
    + 1

Score
/60

I Feel: ☹ 😐 🙂

1. 5
   + 5

2. 7
   + 2

3. 1
   + 6

4. 5
   + 7

5. 9
   + 1

6. 4
   + 5

7. 7
   + 1

8. 1
   + 5

9. 3
   + 7

10. 6
   + 2

11. 5
   + 2

12. 2
   + 1

13. 10
   + 8

14. 2
   + 9

15. 9
   + 6

16. 8
   + 6

17. 6
   + 4

18. 6
   + 8

19. 10
   + 1

20. 9
   + 4

21. 3
   + 4

22. 4
   + 1

23. 6
   + 9

24. 3
   + 10

25. 4
   + 3

26. 10
   + 5

27. 10
   + 3

28. 2
   + 6

29. 7
   + 3

30. 8
   + 5

31. 6
   + 1

32. 8
   + 3

33. 0
   + 3

34. 5
   + 10

35. 2
   + 10

36. 6
   + 2

37. 5
   + 8

38. 3
   + 1

39. 7
   + 9

40. 3
   + 5

41. 8
   + 7

42. 1
   + 9

43. 7
   + 8

44. 8
   + 1

45. 6
   + 7

46. 7
   + 10

47. 6
   + 2

48. 1
   + 7

49. 1
   + 2

50. 3
   + 8

51. 4
   + 8

52. 9
   + 5

53. 10
   + 6

54. 2
   + 7

55. 7
   + 5

56. 6
   + 6

57. 2
   + 8

58. 5
   + 9

59. 1
   + 8

60. 9
   + 8

Score
/60

I Feel:

1.   5
   + 8

2.   6
   + 4

3.   2
   + 7

4.   9
   + 6

5.   10
   + 8

6.   9
   + 5

7.   1
   + 9

8.   1
   + 7

9.   3
   + 8

10.   9
   + 1

11.   6
   + 2

12.   7
   + 5

13.   2
   + 10

14.   7
   + 3

15.   9
   + 8

16.   0
   + 3

17.   3
   + 4

18.   2
   + 8

19.   3
   + 7

20.   6
   + 2

21.   3
   + 1

22.   8
   + 6

23.   10
   + 6

24.   4
   + 1

25.   6
   + 8

26.   5
   + 5

27.   8
   + 3

28.   7
   + 1

29.   10
   + 5

30.   4
   + 3

31.   6
   + 1

32.   4
   + 8

33.   6
   + 6

34.   8
   + 5

35.   8
   + 1

36.   7
   + 2

37.   2
   + 1

38.   8
   + 7

39.   2
   + 6

40.   6
   + 2

41.   1
   + 5

42.   10
   + 3

43.   5
   + 10

44.   3
   + 10

45.   7
   + 8

46.   1
   + 6

47.   7
   + 10

48.   9
   + 4

49.   4
   + 5

50.   5
   + 2

51.   5
   + 7

52.   2
   + 9

53.   5
   + 9

54.   1
   + 8

55.   6
   + 7

56.   6
   + 9

57.   7
   + 9

58.   3
   + 5

59.   1
   + 2

60.   10
   + 1

Score /60

I Feel:

Adding Digits 0 - 10

 Start   Finish

Date: ....   NAME : ....   Time: .....:..... .....:.....

1. 9
    + 5

2. 6
    + 8

3. 1
    + 5

4. 1
    + 7

5. 7
    + 3

6. 6
    + 9

7. 8
    + 3

8. 0
    + 3

9. 8
    + 6

10. 8
     + 1

11. 5
     + 5

12. 10
     + 8

13. 5
     + 9

14. 1
     + 6

15. 8
     + 7

16. 2
     + 1

17. 5
     + 10

18. 3
     + 10

19. 4
     + 8

20. 7
     + 8

21. 4
     + 3

22. 5
     + 8

23. 9
     + 8

24. 2
     + 10

25. 6
     + 6

26. 2
     + 6

27. 2
     + 8

28. 3
     + 5

29. 10
     + 5

30. 6
     + 7

31. 1
     + 2

32. 7
     + 10

33. 7
     + 9

34. 2
     + 9

35. 9
     + 1

36. 3
     + 1

37. 4
     + 5

38. 6
     + 1

39. 7
     + 2

40. 10
     + 3

41. 5
     + 2

42. 6
     + 2

43. 5
     + 7

44. 9
     + 4

45. 3
     + 8

46. 6
     + 4

47. 10
     + 6

48. 8
     + 5

49. 2
     + 7

50. 1
     + 9

51. 1
     + 8

52. 6
     + 2

53. 7
     + 5

54. 7
     + 1

55. 3
     + 7

56. 4
     + 1

57. 9
     + 6

58. 6
     + 2

59. 3
     + 4

60. 10
     + 1

Score /60

I Feel:

Adding Digits 0 - 10

 Start   Finish

Date: ....   NAME : ....   Time: ....:.... ....:....

1.	2.	3.	4.	5.	6.	7.	8.
4 + 8	6 + 8	1 + 8	6 + 2	4 + 5	8 + 3	8 + 1	7 + 9

9.	10.	11.	12.	13.	14.	15.	16.
7 + 8	6 + 4	6 + 2	7 + 5	8 + 6	7 + 3	2 + 9	9 + 4

17.	18.	19.	20.	21.	22.	23.	24.
2 + 6	3 + 4	10 + 1	1 + 6	10 + 5	3 + 10	2 + 8	8 + 7

25.	26.	27.	28.	29.	30.	31.	32.
7 + 2	1 + 7	6 + 6	5 + 8	6 + 2	0 + 3	5 + 7	9 + 5

33.	34.	35.	36.	37.	38.	39.	40.
9 + 1	5 + 2	9 + 6	6 + 9	3 + 5	3 + 1	3 + 8	1 + 9

41.	42.	43.	44.	45.	46.	47.	48.
10 + 8	3 + 7	1 + 2	5 + 5	1 + 5	5 + 10	2 + 1	7 + 10

49.	50.	51.	52.	53.	54.	55.	56.
10 + 6	8 + 5	4 + 3	2 + 10	4 + 1	5 + 9	7 + 1	9 + 8

57.	58.	59.	60.
6 + 1	2 + 7	10 + 3	6 + 7

Score /60

I Feel:

Adding Digits 0 - 10

 Start   Finish

Date: ....   NAME : ....   Time: .....:..... .....:.....

1. 1
    + 5

2. 8
    + 7

3. 4
    + 5

4. 5
    + 2

5. 1
    + 7

6. 0
    + 3

7. 3
    + 0

8. 10
    + 6

9. 6
    + 4

10. 5
    + 0

11. 2
    + 5

12. 9
    + 9

13. 6
    + 0

14. 8
    + 2

15. 4
    + 2

16. 9
    + 4

17. 9
    + 2

18. 3
    + 1

19. 8
    + 1

20. 5
    + 6

21. 3
    + 5

22. 10
    + 2

23. 10
    + 5

24. 8
    + 9

25. 6
    + 7

26. 9
    + 6

27. 3
    + 7

28. 6
    + 9

29. 2
    + 4

30. 7
    + 3

31. 8
    + 6

32. 4
    + 1

33. 7
    + 9

34. 8
    + 5

35. 5
    + 7

36. 9
    + 1

37. 3
    + 9

38. 2
    + 6

39. 5
    + 3

40. 5
    + 9

41. 10
    + 1

42. 4
    + 3

43. 2
    + 7

44. 1
    + 8

45. 9
    + 5

46. 3
    + 10

47. 2
    + 9

48. 0
    + 1

49. 7
    + 0

50. 7
    + 2

51. 2
    + 0

52. 2
    + 1

53. 4
    + 9

54. 5
    + 8

55. 3
    + 6

56. 8
    + 8

57. 9
    + 8

58. 10
    + 0

59. 2
    + 8

60. 1
    + 2

Score /60

I Feel: 🙁 😐 🙂

1.   10   2.   16   3.   15   4.   15   5.   18   6.   15   7.   17   8.   14
   + 15   + 12   + 16   + 19   + 16   + 14   + 16   + 18

9.   17   10.   11   11.   17   12.   12   13.   20   14.   10   15.   19   16.   18
   + 14   + 19   + 15   + 18   + 14   + 17   + 13   + 19

17.   13   18.   11   19.   12   20.   19   21.   17   22.   11   23.   18   24.   14
   + 16   + 12   + 17   + 12   + 19   + 20   + 12   + 17

25.   15   26.   19   27.   16   28.   16   29.   18   30.   14   31.   18   32.   17
   + 20   + 14   + 19   + 13   + 14   + 10   + 13   + 10

33.   16   34.   16   35.   18   36.   18   37.   16   38.   11   39.   14   40.   12
   + 18   + 17   + 11   + 15   + 11   + 16   + 19   + 12

41.   10   42.   18   43.   13   44.   15   45.   19   46.   17   47.   16   48.   13
   + 19   + 18   + 19   + 17   + 20   + 13   + 20   + 10

49.   13   50.   12   51.   19   52.   17   53.   18   54.   12   55.   10   56.   13
   + 14   + 11   + 17   + 18   + 10   + 14   + 18   + 11

57.   19   58.   20   59.   13   60.   14
   + 18   + 19   + 18   + 20

Score /60   I Feel: 😞 😐 🙂

1. 16
    + 18

2. 14
    + 17

3. 17
    + 19

4. 19
    + 14

5. 10
    + 18

6. 11
    + 12

7. 16
    + 17

8. 19
    + 20

9. 18
    + 12

10. 15
    + 14

11. 18
    + 11

12. 18
    + 18

13. 19
    + 18

14. 15
    + 16

15. 17
    + 15

16. 20
    + 19

17. 16
    + 11

18. 18
    + 15

19. 13
    + 11

20. 17
    + 16

21. 10
    + 19

22. 18
    + 14

23. 11
    + 16

24. 14
    + 19

25. 14
    + 18

26. 16
    + 13

27. 13
    + 18

28. 15
    + 17

29. 13
    + 10

30. 13
    + 19

31. 11
    + 19

32. 13
    + 16

33. 16
    + 12

34. 15
    + 19

35. 18
    + 19

36. 12
    + 11

37. 16
    + 20

38. 12
    + 14

39. 20
    + 14

40. 18
    + 13

41. 12
    + 17

42. 19
    + 13

43. 14
    + 10

44. 17
    + 14

45. 17
    + 10

46. 10
    + 15

47. 15
    + 20

48. 18
    + 16

49. 14
    + 20

50. 13
    + 14

51. 16
    + 19

52. 19
    + 17

53. 18
    + 10

54. 10
    + 17

55. 19
    + 12

56. 12
    + 18

57. 17
    + 18

58. 17
    + 13

59. 11
    + 20

60. 12
    + 12

Score
/60

I Feel: ☹ ☺ ☺

1. 11
    + 16

2. 19
    + 17

3. 17
    + 13

4. 12
    + 17

5. 19
    + 12

6. 18
    + 11

7. 18
    + 19

8. 14
    + 20

9. 17
    + 10

10. 14
    + 19

11. 18
    + 18

12. 12
    + 14

13. 17
    + 16

14. 18
    + 15

15. 15
    + 20

16. 18
    + 14

17. 17
    + 18

18. 14
    + 18

19. 19
    + 14

20. 10
    + 18

21. 17
    + 19

22. 12
    + 18

23. 18
    + 13

24. 14
    + 17

25. 20
    + 19

26. 11
    + 19

27. 13
    + 10

28. 18
    + 16

29. 17
    + 14

30. 19
    + 18

31. 11
    + 20

32. 12
    + 12

33. 15
    + 19

34. 14
    + 10

35. 15
    + 16

36. 13
    + 18

37. 16
    + 11

38. 13
    + 14

39. 15
    + 14

40. 10
    + 19

41. 15
    + 17

42. 12
    + 11

43. 18
    + 12

44. 16
    + 12

45. 13
    + 16

46. 10
    + 15

47. 17
    + 15

48. 16
    + 20

49. 16
    + 13

50. 10
    + 17

51. 11
    + 12

52. 13
    + 11

53. 19
    + 13

54. 16
    + 17

55. 13
    + 19

56. 20
    + 14

57. 16
    + 18

58. 18
    + 10

59. 19
    + 20

60. 16
    + 19

Score
/60

I Feel: 😞 😐 😊

Adding Digits 10 - 20

Start   Finish

Date: ....   NAME : ....   Time: .....:..... .....:.....

1. 10 + 17	2. 16 + 20	3. 18 + 19	4. 17 + 15	5. 19 + 12	6. 13 + 11	7. 12 + 17	8. 15 + 16
9. 15 + 20	10. 16 + 13	11. 19 + 18	12. 15 + 19	13. 18 + 16	14. 18 + 14	15. 15 + 14	16. 17 + 14
17. 19 + 14	18. 15 + 17	19. 18 + 12	20. 18 + 11	21. 16 + 17	22. 17 + 19	23. 19 + 17	24. 16 + 11
25. 17 + 10	26. 16 + 18	27. 17 + 13	28. 17 + 18	29. 14 + 19	30. 19 + 20	31. 11 + 20	32. 14 + 20
33. 18 + 13	34. 18 + 18	35. 11 + 19	36. 12 + 18	37. 11 + 16	38. 13 + 14	39. 17 + 16	40. 13 + 10
41. 12 + 11	42. 10 + 19	43. 18 + 15	44. 18 + 10	45. 14 + 17	46. 16 + 19	47. 13 + 18	48. 12 + 12
49. 19 + 13	50. 13 + 16	51. 16 + 12	52. 13 + 19	53. 14 + 10	54. 20 + 19	55. 20 + 14	56. 14 + 18
57. 12 + 14	58. 11 + 12	59. 10 + 15	60. 10 + 18				

Score /60

I Feel: ☹ ☺ ☺

Adding Digits 10 - 20

Date: ....   NAME : ....

   Start   Finish

Time: ....:..... ....:.....

1. 13 + 10	2. 12 + 17	3. 10 + 19	4. 15 + 14	5. 20 + 19	6. 18 + 15	7. 12 + 12	8. 13 + 19
9. 12 + 14	10. 17 + 16	11. 13 + 16	12. 19 + 14	13. 13 + 11	14. 19 + 13	15. 11 + 16	16. 18 + 11
17. 17 + 13	18. 17 + 14	19. 11 + 19	20. 19 + 17	21. 15 + 19	22. 14 + 18	23. 17 + 10	24. 13 + 18
25. 17 + 18	26. 14 + 10	27. 16 + 20	28. 16 + 12	29. 14 + 19	30. 12 + 11	31. 18 + 16	32. 10 + 18
33. 10 + 15	34. 18 + 14	35. 16 + 18	36. 16 + 13	37. 19 + 18	38. 18 + 13	39. 18 + 18	40. 15 + 16
41. 20 + 14	42. 19 + 12	43. 15 + 20	44. 17 + 19	45. 16 + 19	46. 10 + 17	47. 15 + 17	48. 19 + 20
49. 12 + 18	50. 16 + 17	51. 18 + 19	52. 14 + 20	53. 11 + 12	54. 18 + 10	55. 11 + 20	56. 13 + 14
57. 16 + 11	58. 14 + 17	59. 18 + 12	60. 17 + 15				

Score /60

I Feel:

Day 26

Date: ....   NAME : ....

Start   Finish

Time: .....:.....   .....:.....

1. 15 + 16	2. 13 + 14	3. 15 + 19	4. 13 + 16

5. 16
+ 20   6. 14
+ 20   7. 17
+ 14   8. 12
+ 18

9. 11
+ 19   10. 16
+ 12   11. 15
+ 14   12. 18
+ 19   13. 19
+ 18   14. 14
+ 10   15. 10
+ 18   16. 20
+ 19

17. 16
+ 19   18. 12
+ 17   19. 11
+ 20   20. 14
+ 19   21. 15
+ 20   22. 18
+ 18   23. 19
+ 14   24. 13
+ 11

25. 20
+ 14   26. 15
+ 17   27. 10
+ 19   28. 19
+ 13   29. 10
+ 17   30. 19
+ 20   31. 16
+ 18   32. 13
+ 18

33. 14
+ 17   34. 16
+ 13   35. 18
+ 16   36. 17
+ 15   37. 16
+ 17   38. 18
+ 12   39. 18
+ 10   40. 18
+ 15

41. 18
+ 13   42. 18
+ 11   43. 12
+ 12   44. 17
+ 16   45. 11
+ 16   46. 17
+ 13   47. 11
+ 12   48. 17
+ 18

49. 19
+ 17   50. 12
+ 11   51. 16
+ 11   52. 10
+ 15   53. 17
+ 19   54. 12
+ 14   55. 13
+ 19   56. 13
+ 10

57. 18
+ 14   58. 14
+ 18   59. 19
+ 12   60. 17
+ 10

Score /60

I Feel:

1. 12
    + 11

2. 14
    + 20

3. 16
    + 20

4. 16
    + 13

5. 16
    + 11

6. 18
    + 10

7. 14
    + 18

8. 18
    + 14

9. 12
    + 14

10. 10
    + 17

11. 20
    + 19

12. 12
    + 12

13. 17
    + 14

14. 11
    + 12

15. 18
    + 12

16. 18
    + 18

17. 13
    + 18

18. 17
    + 10

19. 19
    + 12

20. 15
    + 20

21. 17
    + 18

22. 10
    + 19

23. 17
    + 19

24. 19
    + 18

25. 18
    + 11

26. 13
    + 19

27. 19
    + 14

28. 19
    + 20

29. 12
    + 17

30. 15
    + 17

31. 15
    + 14

32. 17
    + 15

33. 13
    + 11

34. 18
    + 19

35. 16
    + 12

36. 16
    + 17

37. 14
    + 10

38. 19
    + 13

39. 11
    + 20

40. 18
    + 16

41. 11
    + 19

42. 10
    + 15

43. 14
    + 17

44. 17
    + 16

45. 13
    + 16

46. 12
    + 18

47. 16
    + 18

48. 13
    + 14

49. 18
    + 13

50. 19
    + 17

51. 14
    + 19

52. 11
    + 16

53. 15
    + 19

54. 15
    + 16

55. 17
    + 13

56. 16
    + 19

57. 10
    + 18

58. 13
    + 10

59. 18
    + 15

60. 20
    + 14

Score
/60

I Feel: ☹ 😐 🙂

Adding Digits 10 - 20

Date: ....   NAME : ....

 Start   Finish
Time: .....:.....   .....:.....

1. 19
    + 15

2. 15
    + 18

3. 11
    + 10

4. 18
    + 16

5. 12
    + 17

6. 13
    + 16

7. 14
    + 14

8. 19
    + 11

9. 18
    + 18

10. 18
    + 20

11. 14
    + 17

12. 16
    + 12

13. 11
    + 14

14. 17
    + 19

15. 19
    + 20

16. 13
    + 11

17. 13
    + 15

18. 13
    + 14

19. 19
    + 16

20. 13
    + 17

21. 15
    + 12

22. 16
    + 13

23. 10
    + 16

24. 18
    + 13

25. 16
    + 16

26. 10
    + 15

27. 16
    + 11

28. 20
    + 18

29. 18
    + 15

30. 19
    + 12

31. 11
    + 19

32. 17
    + 20

33. 14
    + 18

34. 17
    + 13

35. 15
    + 19

36. 14
    + 15

37. 14
    + 12

38. 20
    + 16

39. 11
    + 12

40. 17
    + 17

41. 17
    + 16

42. 12
    + 13

43. 15
    + 17

44. 14
    + 16

45. 17
    + 12

46. 19
    + 19

47. 18
    + 10

48. 13
    + 20

49. 16
    + 19

50. 15
    + 13

51. 19
    + 18

52. 10
    + 18

53. 19
    + 13

54. 18
    + 12

55. 13
    + 19

56. 12
    + 15

57. 15
    + 11

58. 11
    + 18

59. 10
    + 17

60. 10
    + 11

Score
/60

I Feel: :(:| :)

Date: .... NAME : .... Time: .....:..... .....:.....

1. 12
    + 9

2. 20
    + 15

3. 3
    + 1

4. 2
    + 15

5. 1
    + 7

6. 3
    + 4

7. 1
    + 3

8. 19
    + 5

9. 5
    + 13

10. 17
    + 1

11. 2
    + 7

12. 8
    + 1

13. 11
    + 9

14. 3
    + 8

15. 11
    + 8

16. 12
    + 8

17. 13
    + 6

18. 12
    + 10

19. 12
    + 2

20. 7
    + 15

21. 18
    + 15

22. 1
    + 9

23. 4
    + 19

24. 14
    + 11

25. 2
    + 13

26. 16
    + 14

27. 9
    + 16

28. 2
    + 5

29. 11
    + 20

30. 19
    + 12

31. 0
    + 10

32. 18
    + 0

33. 1
    + 19

34. 16
    + 17

35. 11
    + 12

36. 8
    + 13

37. 2
    + 8

38. 20
    + 12

39. 6
    + 8

40. 14
    + 10

41. 15
    + 3

42. 6
    + 2

43. 9
    + 3

44. 9
    + 15

45. 3
    + 12

46. 15
    + 1

47. 16
    + 1

48. 18
    + 14

49. 3
    + 20

50. 12
    + 15

51. 8
    + 15

52. 10
    + 17

53. 13
    + 8

54. 8
    + 14

55. 4
    + 12

56. 11
    + 0

57. 10
    + 18

58. 19
    + 14

59. 13
    + 13

60. 15
    + 13

Score /60

I Feel: 😞 😐 😊

Adding Digits 0 - 20

Start   Finish

Date: ....   NAME : ....   Time: .....:.....   .....:.....

1. 8 + 13	2. 18 + 14	3. 11 + 20	4. 4 + 19	5. 12 + 2	6. 13 + 8	7. 2 + 5	8. 19 + 14
9. 11 + 9	10. 1 + 7	11. 2 + 7	12. 19 + 5	13. 15 + 3	14. 15 + 13	15. 13 + 6	16. 9 + 3
17. 5 + 13	18. 16 + 14	19. 11 + 8	20. 3 + 1	21. 14 + 11	22. 2 + 8	23. 8 + 15	24. 20 + 12
25. 15 + 1	26. 7 + 15	27. 6 + 8	28. 13 + 13	29. 11 + 12	30. 1 + 3	31. 20 + 15	32. 18 + 15
33. 4 + 12	34. 10 + 17	35. 2 + 15	36. 8 + 14	37. 0 + 10	38. 3 + 12	39. 3 + 4	40. 1 + 19
41. 12 + 15	42. 1 + 9	43. 3 + 20	44. 9 + 15	45. 12 + 8	46. 12 + 9	47. 2 + 13	48. 10 + 18
49. 6 + 2	50. 11 + 0	51. 18 + 0	52. 17 + 1	53. 16 + 1	54. 19 + 12	55. 3 + 8	56. 16 + 17
57. 8 + 1	58. 12 + 10	59. 14 + 10	60. 9 + 16				

Score /60

I Feel: :(:| :)

Adding Digits 0 - 20

Start   Finish

Date: ....   NAME : ....   Time: .....:..... .....:.....

1. 6 + 8	2. 3 + 1	3. 12 + 9	4. 1 + 3	5. 15 + 13	6. 8 + 15	7. 13 + 13	8. 15 + 1
9. 11 + 0	10. 9 + 15	11. 3 + 4	12. 1 + 19	13. 8 + 13	14. 18 + 0	15. 3 + 20	16. 16 + 14
17. 20 + 12	18. 13 + 6	19. 19 + 14	20. 9 + 3	21. 8 + 1	22. 3 + 8	23. 5 + 13	24. 2 + 8
25. 3 + 12	26. 2 + 5	27. 2 + 7	28. 6 + 2	29. 11 + 12	30. 10 + 17	31. 9 + 16	32. 11 + 20
33. 19 + 12	34. 10 + 18	35. 18 + 14	36. 15 + 3	37. 1 + 7	38. 14 + 11	39. 12 + 15	40. 12 + 10
41. 4 + 19	42. 13 + 8	43. 19 + 5	44. 20 + 15	45. 16 + 17	46. 8 + 14	47. 12 + 8	48. 11 + 9
49. 1 + 9	50. 11 + 8	51. 4 + 12	52. 0 + 10	53. 12 + 2	54. 16 + 1	55. 17 + 1	56. 7 + 15
57. 18 + 15	58. 2 + 13	59. 2 + 15	60. 14 + 10				

Score /60   I Feel:

1. 8 + 13
2. 20 + 12
3. 9 + 16
4. 6 + 8
5. 1 + 7
6. 0 + 10
7. 19 + 5
8. 15 + 13

9. 6 + 2
10. 7 + 15
11. 14 + 11
12. 2 + 7
13. 16 + 14
14. 13 + 8
15. 2 + 5
16. 13 + 6

17. 19 + 14
18. 18 + 15
19. 2 + 13
20. 15 + 3
21. 20 + 15
22. 8 + 14
23. 9 + 15
24. 16 + 1

25. 11 + 8
26. 11 + 9
27. 3 + 1
28. 3 + 12
29. 3 + 4
30. 1 + 19
31. 12 + 8
32. 17 + 1

33. 18 + 0
34. 16 + 17
35. 12 + 2
36. 14 + 10
37. 4 + 19
38. 12 + 15
39. 2 + 8
40. 3 + 20

41. 10 + 18
42. 9 + 3
43. 10 + 17
44. 1 + 3
45. 13 + 13
46. 8 + 15
47. 11 + 12
48. 18 + 14

49. 1 + 9
50. 2 + 15
51. 15 + 1
52. 11 + 0
53. 19 + 12
54. 12 + 9
55. 4 + 12
56. 3 + 8

57. 11 + 20
58. 12 + 10
59. 8 + 1
60. 5 + 13

Score /60

I Feel: 😟 😐 🙂

Day 33

Date: ....   NAME : ....

Start   Finish

Time: ....:..... ....:.....

1. 12 + 14	2. 15 + 9	3. 14 + 4	4. 7 + 19

1. 12
   + 14

2. 15
   + 9

3. 14
   + 4

4. 7
   + 19

5. 13
   + 4

6. 9
   + 7

7. 14
   + 15

8. 10
   + 1

9. 12
   + 19

10. 11
    + 12

11. 17
    + 16

12. 7
    + 6

13. 16
    + 13

14. 1
    + 9

15. 9
    + 10

16. 19
    + 12

17. 8
    + 13

18. 6
    + 5

19. 19
    + 13

20. 10
    + 16

21. 6
    + 14

22. 8
    + 10

23. 5
    + 19

24. 16
    + 7

25. 4
    + 1

26. 9
    + 6

27. 1
    + 17

28. 12
    + 3

29. 5
    + 2

30. 12
    + 0

31. 7
    + 6

32. 5
    + 10

33. 6
    + 13

34. 7
    + 20

35. 16
    + 10

36. 6
    + 2

37. 20
    + 17

38. 15
    + 4

39. 9
    + 1

40. 2
    + 2

41. 7
    + 5

42. 15
    + 19

43. 9
    + 12

44. 12
    + 2

45. 3
    + 17

46. 9
    + 14

47. 4
    + 9

48. 1
    + 15

49. 10
    + 2

50. 18
    + 13

51. 13
    + 2

52. 8
    + 3

53. 15
    + 14

54. 17
    + 15

55. 3
    + 2

56. 11
    + 1

57. 18
    + 18

58. 17
    + 4

59. 16
    + 3

60. 6
    + 10

Score /60

I Feel: 😞 😐 😊

1. 6
   + 14

2. 9
   + 1

3. 4
   + 9

4. 19
   + 12

5. 4
   + 1

6. 17
   + 15

7. 15
   + 19

8. 8
   + 13

9. 3
   + 17

10. 11
    + 1

11. 14
    + 4

12. 13
    + 2

13. 5
    + 19

14. 16
    + 13

15. 13
    + 4

16. 12
    + 2

17. 9
   + 14

18. 7
   + 6

19. 3
   + 2

20. 17
   + 4

21. 16
   + 7

22. 6
   + 13

23. 17
   + 16

24. 9
   + 10

25. 12
   + 0

26. 20
   + 17

27. 11
   + 12

28. 18
   + 13

29. 9
   + 6

30. 16
   + 3

31. 5
   + 2

32. 9
   + 7

33. 6
   + 5

34. 1
   + 15

35. 14
   + 15

36. 2
   + 2

37. 19
   + 13

38. 5
   + 10

39. 9
   + 12

40. 6
   + 10

41. 8
   + 10

42. 1
   + 9

43. 15
   + 14

44. 7
   + 5

45. 1
   + 17

46. 18
   + 18

47. 7
   + 19

48. 10
   + 1

49. 10
   + 2

50. 12
   + 14

51. 16
   + 10

52. 15
   + 4

53. 7
   + 6

54. 6
   + 2

55. 15
   + 9

56. 7
   + 20

57. 10
   + 16

58. 12
   + 3

59. 8
   + 3

60. 12
   + 19

Score
/60

I Feel: :(:| :)

1. 9
   + 1

2. 16
   + 7

3. 8
   + 3

4. 19
   + 12

5. 9
   + 10

6. 7
   + 6

7. 10
   + 1

8. 1
   + 9

9. 8
   + 13

10. 10
   + 16

11. 5
   + 10

12. 13
   + 2

13. 15
   + 4

14. 12
   + 0

15. 16
   + 13

16. 7
   + 6

17. 19
   + 13

18. 9
   + 14

19. 12
   + 19

20. 1
   + 17

21. 9
   + 12

22. 15
   + 19

23. 4
   + 1

24. 11
   + 1

25. 17
   + 16

26. 16
   + 3

27. 18
   + 18

28. 7
   + 19

29. 13
   + 4

30. 9
   + 7

31. 5
   + 19

32. 16
   + 10

33. 18
   + 13

34. 5
   + 2

35. 14
   + 15

36. 11
   + 12

37. 15
   + 14

38. 3
   + 2

39. 1
   + 15

40. 3
   + 17

41. 12
   + 14

42. 6
   + 2

43. 14
   + 4

44. 6
   + 5

45. 7
   + 20

46. 12
   + 2

47. 4
   + 9

48. 17
   + 15

49. 15
   + 9

50. 10
   + 2

51. 6
   + 14

52. 17
   + 4

53. 6
   + 13

54. 12
   + 3

55. 2
   + 2

56. 20
   + 17

57. 6
   + 10

58. 7
   + 5

59. 8
   + 10

60. 9
   + 6

Score /60

I Feel:

1. 1
    + 17

2. 6
    + 5

3. 8
    + 13

4. 6
    + 2

5. 17
    + 16

6. 10
    + 16

7. 7
    + 6

8. 12
    + 19

9. 10
    + 1

10. 11
    + 1

11. 15
    + 14

12. 15
    + 19

13. 5
    + 2

14. 8
    + 10

15. 7
    + 20

16. 3
    + 2

17. 18
    + 18

18. 16
    + 10

19. 7
    + 5

20. 9
    + 1

21. 13
    + 2

22. 8
    + 3

23. 6
    + 14

24. 9
    + 10

25. 9
    + 12

26. 10
    + 2

27. 14
    + 4

28. 9
    + 6

29. 9
    + 14

30. 15
    + 4

31. 3
    + 17

32. 9
    + 7

33. 12
    + 0

34. 12
    + 3

35. 13
    + 4

36. 5
    + 10

37. 1
    + 9

38. 19
    + 12

39. 6
    + 13

40. 4
    + 9

41. 7
    + 19

42. 2
    + 2

43. 6
    + 10

44. 17
    + 15

45. 20
    + 17

46. 14
    + 15

47. 7
    + 6

48. 1
    + 15

49. 5
    + 19

50. 12
    + 2

51. 12
    + 14

52. 16
    + 13

53. 15
    + 9

54. 4
    + 1

55. 16
    + 3

56. 11
    + 12

57. 18
    + 13

58. 16
    + 7

59. 19
    + 13

60. 17
    + 4

Score
/60

I Feel: ☹ ☺ 😊

Day 37

Date: ....

NAME : ....

Start   Finish

Time: .....:.....   .....:.....

1.	2.	3.	4.	5.	6.	7.	8.
7 + 6	7 + 19	8 + 10	19 + 12	9 + 14	1 + 9	16 + 3	15 + 4

9.	10.	11.	12.	13.	14.	15.	16.
13 + 4	18 + 18	7 + 6	3 + 2	5 + 2	11 + 1	16 + 10	7 + 5

17.	18.	19.	20.	21.	22.	23.	24.
10 + 1	16 + 13	7 + 20	1 + 15	14 + 15	4 + 1	12 + 0	9 + 1

25.	26.	27.	28.	29.	30.	31.	32.
1 + 17	6 + 5	11 + 12	3 + 17	12 + 3	6 + 14	9 + 6	9 + 12

33.	34.	35.	36.	37.	38.	39.	40.
13 + 2	20 + 17	2 + 2	18 + 13	15 + 14	4 + 9	17 + 16	12 + 14

41.	42.	43.	44.	45.	46.	47.	48.
12 + 2	8 + 3	17 + 4	9 + 10	17 + 15	14 + 4	6 + 13	16 + 7

49.	50.	51.	52.	53.	54.	55.	56.
5 + 19	10 + 16	8 + 13	15 + 19	6 + 10	10 + 2	6 + 2	19 + 13

57.	58.	59.	60.
9 + 7	12 + 19	15 + 9	5 + 10

Score
/60

I Feel:

Adding Digits 0 - 20

Date: ....   NAME : ....

    Start   Finish

Time: .....:.....   .....:.....

1. 11 + 12	2. 16 + 7	3. 9 + 1	4. 12 + 3	5. 1 + 15	6. 18 + 13	7. 9 + 10	8. 5 + 19
9. 15 + 4	10. 4 + 1	11. 6 + 10	12. 8 + 13	13. 12 + 0	14. 6 + 14	15. 3 + 2	16. 18 + 18
17. 10 + 2	18. 5 + 10	19. 14 + 15	20. 13 + 4	21. 6 + 5	22. 12 + 14	23. 6 + 13	24. 6 + 2
25. 17 + 16	26. 19 + 12	27. 15 + 19	28. 15 + 14	29. 9 + 7	30. 2 + 2	31. 8 + 10	32. 20 + 17
33. 4 + 9	34. 10 + 1	35. 17 + 15	36. 15 + 9	37. 10 + 16	38. 12 + 2	39. 12 + 19	40. 14 + 4
41. 17 + 4	42. 9 + 12	43. 5 + 2	44. 1 + 9	45. 9 + 6	46. 7 + 5	47. 1 + 17	48. 7 + 19
49. 16 + 13	50. 19 + 13	51. 8 + 3	52. 7 + 6	53. 16 + 10	54. 7 + 6	55. 16 + 3	56. 3 + 17
57. 13 + 2	58. 11 + 1	59. 7 + 20	60. 9 + 14				

Score /60

I Feel:

Adding Digits 0 - 20

 Start   Finish

Date: ....   NAME : ....   Time: .....:..... .....:.....

1. 5 + 2	2. 12 + 3	3. 16 + 10	4. 9 + 7	5. 14 + 15	6. 12 + 0	7. 4 + 1	8. 11 + 12
9. 1 + 9	10. 5 + 10	11. 6 + 5	12. 9 + 10	13. 9 + 14	14. 4 + 9	15. 13 + 4	16. 14 + 4
17. 12 + 14	18. 10 + 16	19. 18 + 13	20. 6 + 14	21. 8 + 10	22. 16 + 13	23. 15 + 4	24. 11 + 1
25. 7 + 5	26. 16 + 7	27. 5 + 19	28. 19 + 12	29. 3 + 2	30. 7 + 20	31. 13 + 2	32. 15 + 14
33. 16 + 3	34. 1 + 17	35. 10 + 2	36. 8 + 3	37. 7 + 6	38. 19 + 13	39. 7 + 6	40. 10 + 1
41. 6 + 2	42. 12 + 2	43. 15 + 19	44. 18 + 18	45. 12 + 19	46. 7 + 19	47. 9 + 1	48. 1 + 15
49. 6 + 10	50. 15 + 9	51. 17 + 15	52. 2 + 2	53. 20 + 17	54. 9 + 12	55. 9 + 6	56. 3 + 17
57. 17 + 4	58. 17 + 16	59. 6 + 13	60. 8 + 13				

Score /60   I Feel: 😟 😐 🙂

Adding Digits 0 - 20

Start   Finish

Date: ....   NAME : ....   Time: ......:......   ......:......

1.	2.	3.	4.	5.	6.	7.	8.
6 + 2	9 + 7	8 + 10	2 + 2	17 + 15	1 + 17	9 + 12	15 + 9

9.	10.	11.	12.	13.	14.	15.	16.
1 + 9	3 + 2	5 + 10	19 + 13	9 + 6	12 + 0	13 + 4	5 + 19

17.	18.	19.	20.	21.	22.	23.	24.
14 + 4	10 + 1	8 + 13	10 + 2	14 + 15	11 + 12	7 + 6	9 + 10

25.	26.	27.	28.	29.	30.	31.	32.
12 + 19	10 + 16	5 + 2	7 + 5	7 + 19	9 + 14	4 + 1	6 + 13

33.	34.	35.	36.	37.	38.	39.	40.
6 + 5	1 + 15	16 + 3	4 + 9	8 + 3	11 + 1	3 + 17	7 + 6

41.	42.	43.	44.	45.	46.	47.	48.
15 + 4	18 + 18	9 + 1	20 + 17	12 + 14	18 + 13	15 + 19	6 + 14

49.	50.	51.	52.	53.	54.	55.	56.
16 + 10	17 + 16	6 + 10	19 + 12	12 + 3	16 + 7	16 + 13	7 + 20

57.	58.	59.	60.
12 + 2	13 + 2	15 + 14	17 + 4

Score /60

I Feel:

Subtracting Digits 0 - 5

Date: ....   NAME : ....

   Start   Finish

Time: .....:.....   .....:.....

1. 3 - 2	2. 4 - 3	3. 4 - 1	4. 2 - 1	5. 4 - 2	6. 4 - 0	7. 3 - 0	8. 5 - 1

1.   3
  - 2

2.   4
  - 3

3.   4
  - 1

4.   2
  - 1

5.   4
  - 2

6.   4
  - 0

7.   3
  - 0

8.   5
  - 1

9.   2
  - 0

10.   5
  - 2

11.   3
  - 1

12.   5
  - 0

13.   5
  - 3

14.   5
  - 4

15.   5
  - 1

16.   2
  - 0

17.   4
  - 3

18.   4
  - 1

19.   3
  - 2

20.   3
  - 1

21.   5
  - 1

22.   3
  - 2

23.   3
  - 1

24.   2
  - 0

25.   3
  - 1

26.   4
  - 0

27.   2
  - 1

28.   5
  - 1

29.   4
  - 1

30.   3
  - 0

31.   3
  - 2

32.   5
  - 3

33.   4
  - 1

34.   5
  - 0

35.   4
  - 1

36.   4
  - 1

37.   3
  - 2

38.   3
  - 1

39.   3
  - 0

40.   5
  - 2

41.   4
  - 0

42.   3
  - 1

43.   4
  - 2

44.   5
  - 1

45.   2
  - 1

46.   4
  - 0

47.   2
  - 1

48.   4
  - 3

49.   2
  - 1

50.   3
  - 2

51.   2
  - 0

52.   2
  - 0

53.   3
  - 0

54.   3
  - 0

55.   2
  - 0

56.   3
  - 0

57.   2
  - 0

58.   3
  - 0

59.   2
  - 1

60.   2
  - 1

Score /60

I Feel: 😟 😐 🙂

1. 3
    - 0

2. 3
    - 1

3. 4
    - 1

4. 4
    - 2

5. 5
    - 0

6. 5
    - 3

7. 3
    - 0

8. 3
    - 2

9. 4
    - 2

10. 4
    - 1

11. 2
    - 1

12. 2
    - 0

13. 3
    - 2

14. 3
    - 2

15. 2
    - 1

16. 3
    - 0

17. 5
    - 1

18. 5
    - 4

19. 4
    - 1

20. 5
    - 3

21. 5
    - 4

22. 4
    - 3

23. 4
    - 3

24. 4
    - 3

25. 4
    - 2

26. 4
    - 0

27. 3
    - 2

28. 5
    - 2

29. 4
    - 1

30. 2
    - 1

31. 4
    - 2

32. 2
    - 0

33. 4
    - 2

34. 2
    - 0

35. 4
    - 1

36. 3
    - 0

37. 4
    - 3

38. 3
    - 1

39. 4
    - 3

40. 3
    - 0

41. 4
    - 1

42. 5
    - 4

43. 4
    - 3

44. 5
    - 0

45. 5
    - 0

46. 4
    - 3

47. 3
    - 1

48. 2
    - 0

49. 3
    - 2

50. 5
    - 2

51. 3
    - 2

52. 3
    - 1

53. 4
    - 2

54. 3
    - 2

55. 3
    - 2

56. 5
    - 3

57. 5
    - 1

58. 3
    - 1

59. 2
    - 0

60. 5
    - 4

Score /60

I Feel:

Day 43

Date: ....

NAME : ....

Start   Finish

Time: ......:...... ......:......

| 1. | 4
- 3 | 2. | 5
- 2 | 3. | 2
- 1 | 4. | 2
- 0 | 5. | 4
- 1 | 6. | 4
- 0 | 7. | 2
- 2 | 8. | 3
- 2 |

| 9. | 2
- 1 | 10. | 3
- 0 | 11. | 5
- 3 | 12. | 3
- 0 | 13. | 4
- 1 | 14. | 5
- 0 | 15. | 4
- 2 | 16. | 3
- 1 |

| 17. | 3
- 2 | 18. | 3
- 2 | 19. | 2
- 0 | 20. | 3
- 2 | 21. | 4
- 1 | 22. | 4
- 2 | 23. | 4
- 3 | 24. | 4
- 2 |

| 25. | 5
- 3 | 26. | 2
- 0 | 27. | 3
- 1 | 28. | 4
- 1 | 29. | 3
- 1 | 30. | 5
- 4 | 31. | 3
- 0 | 32. | 2
- 1 |

| 33. | 4
- 3 | 34. | 5
- 4 | 35. | 4
- 3 | 36. | 4
- 2 | 37. | 3
- 3 | 38. | 3
- 2 | 39. | 5
- 4 | 40. | 4
- 1 |

| 41. | 4
- 1 | 42. | 5
- 1 | 43. | 5
- 1 | 44. | 4
- 2 | 45. | 5
- 0 | 46. | 5
- 3 | 47. | 5
- 2 | 48. | 3
- 1 |

| 49. | 3
- 0 | 50. | 5
- 0 | 51. | 3
- 0 | 52. | 4
- 3 | 53. | 2
- 0 | 54. | 4
- 2 | 55. | 3
- 2 | 56. | 5
- 4 |

| 57. | 2
- 0 | 58. | 3
- 1 | 59. | 3
- 2 | 60. | 3
- 2 |

Score /60

I Feel:

Subtracting Digits 0 - 5

 Start   Finish

Date: ....   NAME : ....   Time: .....:..... .....:.....

1. 4 − 2

2. 5 − 2

3. 4 − 2

4. 3 − 3

5. 5 − 0

6. 5 − 2

7. 3 − 2

8. 2 − 0

9. 3 − 0

10. 3 − 2

11. 5 − 4

12. 3 − 1

13. 5 − 4

14. 3 − 1

15. 2 − 0

16. 2 − 1

17. 4 − 0

18. 5 − 1

19. 2 − 2

20. 4 − 1

21. 3 − 1

22. 3 − 0

23. 3 − 2

24. 3 − 2

25. 5 − 4

26. 5 − 3

27. 3 − 2

28. 4 − 3

29. 2 − 0

30. 4 − 3

31. 5 − 1

32. 5 − 0

33. 3 − 2

34. 4 − 1

35. 3 − 0

36. 2 − 1

37. 5 − 3

38. 3 − 0

39. 4 − 1

40. 4 − 2

41. 3 − 0

42. 3 − 2

43. 4 − 3

44. 4 − 1

45. 4 − 2

46. 3 − 1

47. 5 − 0

48. 3 − 1

49. 2 − 0

50. 4 − 2

51. 4 − 2

52. 5 − 3

53. 4 − 1

54. 5 − 4

55. 4 − 1

56. 3 − 2

57. 2 − 1

58. 2 − 0

59. 4 − 3

60. 4 − 3

Score /60

I Feel: ☹ 😐 ☺

1. 2
  − 0

2. 3
  − 1

3. 3
  − 2

4. 5
  − 3

5. 6
  − 1

6. 3
  − 0

7. 6
  − 4

8. 6
  − 2

9. 6
  − 5

10. 6
  − 5

11. 7
  − 4

12. 4
  − 3

13. 7
  − 4

14. 2
  − 1

15. 3
  − 1

16. 6
  − 1

17. 7
  − 1

18. 3
  − 0

19. 7
  − 1

20. 2
  − 1

21. 7
  − 5

22. 5
  − 4

23. 7
  − 2

24. 4
  − 2

25. 2
  − 0

26. 6
  − 0

27. 6
  − 2

28. 5
  − 1

29. 5
  − 3

30. 4
  − 3

31. 6
  − 1

32. 6
  − 1

33. 4
  − 2

34. 3
  − 2

35. 4
  − 3

36. 6
  − 3

37. 4
  − 2

38. 5
  − 0

39. 7
  − 3

40. 4
  − 1

41. 6
  − 2

42. 3
  − 0

43. 4
  − 2

44. 3
  − 0

45. 7
  − 6

46. 2
  − 0

47. 5
  − 2

48. 5
  − 1

49. 5
  − 1

50. 4
  − 2

51. 5
  − 3

52. 6
  − 2

53. 4
  − 0

54. 2
  − 0

55. 4
  − 0

56. 7
  − 0

57. 6
  − 4

58. 2
  − 0

59. 5
  − 0

60. 7
  − 1

Score
/60

I Feel: 😕 😐 🙂

Subtracting Digits 0 - 7

Date: ....   NAME : ....    Start   Finish

Time: .....:.....   .....:.....

1. $\begin{array}{r} 6 \\ -5 \end{array}$	2. $\begin{array}{r} 7 \\ -2 \end{array}$	3. $\begin{array}{r} 4 \\ -2 \end{array}$	4. $\begin{array}{r} 2 \\ -1 \end{array}$	5. $\begin{array}{r} 3 \\ -1 \end{array}$	6. $\begin{array}{r} 4 \\ -1 \end{array}$	7. $\begin{array}{r} 6 \\ -0 \end{array}$	8. $\begin{array}{r} 3 \\ -2 \end{array}$
9. $\begin{array}{r} 5 \\ -3 \end{array}$	10. $\begin{array}{r} 7 \\ -3 \end{array}$	11. $\begin{array}{r} 4 \\ -0 \end{array}$	12. $\begin{array}{r} 3 \\ -0 \end{array}$	13. $\begin{array}{r} 4 \\ -3 \end{array}$	14. $\begin{array}{r} 7 \\ -4 \end{array}$	15. $\begin{array}{r} 2 \\ -0 \end{array}$	16. $\begin{array}{r} 6 \\ -2 \end{array}$
17. $\begin{array}{r} 7 \\ -0 \end{array}$	18. $\begin{array}{r} 7 \\ -5 \end{array}$	19. $\begin{array}{r} 5 \\ -4 \end{array}$	20. $\begin{array}{r} 6 \\ -3 \end{array}$	21. $\begin{array}{r} 5 \\ -2 \end{array}$	22. $\begin{array}{r} 6 \\ -4 \end{array}$	23. $\begin{array}{r} 5 \\ -1 \end{array}$	24. $\begin{array}{r} 6 \\ -1 \end{array}$
25. $\begin{array}{r} 7 \\ -6 \end{array}$	26. $\begin{array}{r} 7 \\ -1 \end{array}$	27. $\begin{array}{r} 5 \\ -0 \end{array}$	28. $\begin{array}{r} 7 \\ -5 \end{array}$	29. $\begin{array}{r} 4 \\ -1 \end{array}$	30. $\begin{array}{r} 5 \\ -3 \end{array}$	31. $\begin{array}{r} 2 \\ -1 \end{array}$	32. $\begin{array}{r} 4 \\ -2 \end{array}$
33. $\begin{array}{r} 4 \\ -3 \end{array}$	34. $\begin{array}{r} 3 \\ -2 \end{array}$	35. $\begin{array}{r} 6 \\ -0 \end{array}$	36. $\begin{array}{r} 3 \\ -2 \end{array}$	37. $\begin{array}{r} 4 \\ -0 \end{array}$	38. $\begin{array}{r} 2 \\ -1 \end{array}$	39. $\begin{array}{r} 4 \\ -3 \end{array}$	40. $\begin{array}{r} 5 \\ -3 \end{array}$
41. $\begin{array}{r} 7 \\ -6 \end{array}$	42. $\begin{array}{r} 7 \\ -2 \end{array}$	43. $\begin{array}{r} 5 \\ -3 \end{array}$	44. $\begin{array}{r} 6 \\ -5 \end{array}$	45. $\begin{array}{r} 5 \\ -3 \end{array}$	46. $\begin{array}{r} 3 \\ -2 \end{array}$	47. $\begin{array}{r} 7 \\ -4 \end{array}$	48. $\begin{array}{r} 3 \\ -3 \end{array}$
49. $\begin{array}{r} 6 \\ -3 \end{array}$	50. $\begin{array}{r} 7 \\ -1 \end{array}$	51. $\begin{array}{r} 3 \\ -0 \end{array}$	52. $\begin{array}{r} 4 \\ -1 \end{array}$	53. $\begin{array}{r} 2 \\ -1 \end{array}$	54. $\begin{array}{r} 4 \\ -3 \end{array}$	55. $\begin{array}{r} 6 \\ -3 \end{array}$	56. $\begin{array}{r} 3 \\ -0 \end{array}$
57. $\begin{array}{r} 5 \\ -3 \end{array}$	58. $\begin{array}{r} 5 \\ -2 \end{array}$	59. $\begin{array}{r} 6 \\ -4 \end{array}$	60. $\begin{array}{r} 3 \\ -1 \end{array}$				

Score /60

I Feel: 😦 😐 🙂

Subtracting Digits 0 - 7

Date: ....

NAME : ....

Start   Finish

Time: ....:..... ....:.....

1. 6 − 2	2. 7 − 1	3. 4 − 3	4. 4 − 2	5. 7 − 2	6. 3 − 2	7. 6 − 0	8. 4 − 0
9. 2 − 0	10. 3 − 1	11. 7 − 0	12. 6 − 2	13. 7 − 4	14. 5 − 3	15. 6 − 2	16. 6 − 4
17. 6 − 1	18. 6 − 2	19. 5 − 4	20. 7 − 1	21. 5 − 3	22. 4 − 2	23. 2 − 0	24. 2 − 0
25. 5 − 3	26. 4 − 2	27. 3 − 0	28. 3 − 2	29. 4 − 3	30. 2 − 1	31. 6 − 1	32. 6 − 3
33. 2 − 1	34. 5 − 2	35. 7 − 5	36. 2 − 0	37. 2 − 0	38. 6 − 4	39. 5 − 1	40. 4 − 2
41. 4 − 1	42. 4 − 0	43. 6 − 1	44. 7 − 6	45. 3 − 1	46. 6 − 5	47. 7 − 1	48. 4 − 2
49. 3 − 0	50. 3 − 0	51. 7 − 4	52. 5 − 0	53. 7 − 3	54. 6 − 5	55. 5 − 0	56. 5 − 1
57. 5 − 1	58. 6 − 1	59. 3 − 0	60. 4 − 3				

Score /60

I Feel:

1. 3 - 0	2. 6 - 3	3. 6 - 0	4. 7 - 2	5. 2 - 0	6. 3 - 2	7. 6 - 4	8. 3 - 1

9. 4 - 3	10. 7 - 1	11. 4 - 2	12. 7 - 5	13. 7 - 6	14. 4 - 2	15. 5 - 3	16. 7 - 1

17. 7 - 6	18. 2 - 1	19. 7 - 2	20. 5 - 3	21. 7 - 4	22. 3 - 2	23. 6 - 3	24. 2 - 1

25. 6 - 0	26. 3 - 2	27. 3 - 1	28. 5 - 3	29. 3 - 0	30. 6 - 1	31. 3 - 0	32. 5 - 0

33. 4 - 1	34. 4 - 3	35. 6 - 3	36. 4 - 3	37. 6 - 4	38. 5 - 2	39. 3 - 2	40. 7 - 3

41. 5 - 4	42. 5 - 3	43. 5 - 3	44. 2 - 1	45. 4 - 1	46. 7 - 5	47. 3 - 3	48. 5 - 1

49. 7 - 0	50. 6 - 5	51. 2 - 1	52. 6 - 2	53. 7 - 4	54. 6 - 5	55. 4 - 1	56. 4 - 0

57. 4 - 3	58. 5 - 2	59. 4 - 0	60. 5 - 3

Score
/60

I Feel:

1. 7
  - 2

2. 5
  - 3

3. 2
  - 1

4. 4
  - 2

5. 6
  - 1

6. 6
  - 3

7. 5
  - 3

8. 7
  - 0

9. 5
  - 2

10. 3
   - 0

11. 6
   - 3

12. 3
   - 2

13. 4
   - 0

14. 4
   - 3

15. 7
   - 6

16. 7
   - 1

17. 5
   - 4

18. 3
   - 3

19. 5
   - 3

20. 4
   - 3

21. 6
   - 2

22. 4
   - 3

23. 5
   - 3

24. 3
   - 2

25. 6
   - 4

26. 5
   - 3

27. 7
   - 1

28. 5
   - 2

29. 6
   - 0

30. 7
   - 4

31. 4
   - 1

32. 7
   - 4

33. 4
   - 1

34. 6
   - 4

35. 6
   - 0

36. 3
   - 1

37. 7
   - 5

38. 4
   - 3

39. 2
   - 0

40. 5
   - 0

41. 3
   - 0

42. 4
   - 0

43. 6
   - 5

44. 5
   - 1

45. 3
   - 2

46. 3
   - 2

47. 2
   - 1

48. 7
   - 3

49. 3
   - 1

50. 7
   - 5

51. 2
   - 1

52. 2
   - 1

53. 6
   - 3

54. 7
   - 6

55. 6
   - 5

56. 3
   - 0

57. 4
   - 1

58. 5
   - 3

59. 7
   - 2

60. 4
   - 2

Score /60

I Feel:

Day 50

Subtracting Digits 0 - 7

 Start   Finish

Date: ....   NAME : ....   Time: .....:..... .....:.....

1. 3 − 2	2. 5 − 3	3. 5 − 2	4. 4 − 1	5. 5 − 3	6. 3 − 2	7. 5 − 3	8. 6 − 0
9. 7 − 1	10. 6 − 1	11. 5 − 0	12. 4 − 0	13. 5 − 2	14. 5 − 3	15. 4 − 3	16. 6 − 5
17. 3 − 1	18. 5 − 4	19. 4 − 0	20. 6 − 4	21. 5 − 1	22. 6 − 3	23. 6 − 6	24. 7 − 2
25. 4 − 1	26. 7 − 6	27. 2 − 1	28. 3 − 0	29. 4 − 3	30. 7 − 4	31. 3 − 0	32. 7 − 1
33. 4 − 2	34. 6 − 4	35. 7 − 2	36. 2 − 1	37. 5 − 3	38. 5 − 2	39. 4 − 3	40. 7 − 5
41. 7 − 0	42. 7 − 3	43. 3 − 3	44. 6 − 3	45. 5 − 3	46. 4 − 2	47. 6 − 4	48. 7 − 6
49. 6 − 2	50. 7 − 5	51. 2 − 1	52. 4 − 3	53. 2 − 1	54. 3 − 1	55. 6 − 0	56. 3 − 0
57. 3 − 2	58. 2 − 0	59. 6 − 5	60. 7 − 4				

Score /60   I Feel:

1. 8
    − 6

2. 5
    − 1

3. 10
    − 3

4. 8
    − 0

5. 7
    − 3

6. 6
    − 1

7. 10
    − 8

8. 5
    − 2

9. 2
    − 0

10. 10
    − 2

11. 3
    − 0

12. 7
    − 4

13. 7
    − 0

14. 9
    − 1

15. 9
    − 8

16. 3
    − 1

17. 4
    − 0

18. 6
    − 2

19. 9
    − 4

20. 6
    − 3

21. 7
    − 2

22. 10
    − 6

23. 4
    − 3

24. 6
    − 0

25. 10
    − 4

26. 8
    − 3

27. 5
    − 2

28. 9
    − 3

29. 7
    − 6

30. 6
    − 4

31. 7
    − 0

32. 4
    − 1

33. 10
    − 7

34. 7
    − 1

35. 9
    − 7

36. 8
    − 3

37. 8
    − 7

38. 10
    − 1

39. 9
    − 5

40. 5
    − 0

41. 10
    − 5

42. 8
    − 1

43. 3
    − 2

44. 5
    − 3

45. 4
    − 3

46. 9
    − 6

47. 4
    − 2

48. 7
    − 6

49. 5
    − 4

50. 9
    − 5

51. 3
    − 1

52. 7
    − 5

53. 2
    − 1

54. 9
    − 2

55. 8
    − 4

56. 10
    − 9

57. 8
    − 2

58. 6
    − 5

59. 9
    − 0

60. 8
    − 5

Score /60   I Feel:

1. 6
    − 3

2. 3
    − 0

3. 10
    − 9

4. 7
    − 6

5. 8
    − 2

6. 3
    − 1

7. 7
    − 2

8. 10
    − 5

9. 9
    − 2

10. 10
    − 7

11. 2
    − 1

12. 8
    − 5

13. 10
    − 4

14. 10
    − 8

15. 5
    − 3

16. 3
    − 1

17. 9
    − 0

18. 10
    − 6

19. 9
    − 8

20. 7
    − 6

21. 9
    − 7

22. 2
    − 0

23. 4
    − 3

24. 10
    − 3

25. 7
    − 0

26. 8
    − 0

27. 9
    − 5

28. 4
    − 3

29. 3
    − 2

30. 7
    − 0

31. 9
    − 6

32. 7
    − 4

33. 4
    − 0

34. 8
    − 4

35. 5
    − 2

36. 8
    − 6

37. 6
    − 2

38. 9
    − 1

39. 9
    − 4

40. 10
    − 1

41. 5
    − 4

42. 8
    − 3

43. 6
    − 5

44. 7
    − 1

45. 4
    − 1

46. 5
    − 2

47. 9
    − 3

48. 8
    − 7

49. 10
    − 2

50. 7
    − 3

51. 6
    − 0

52. 7
    − 5

53. 9
    − 5

54. 5
    − 1

55. 8
    − 3

56. 6
    − 1

57. 5
    − 0

58. 4
    − 2

59. 8
    − 1

60. 6
    − 4

Score /60

I Feel: ☹ 😐 🙂

1.   2
   - 0

2.   5
   - 0

3.   7
   - 5

4.   9
   - 4

5.   6
   - 5

6.   5
   - 4

7.   8
   - 3

8.   8
   - 0

9.   4
   - 2

10. 7
   - 1

11. 9
   - 1

12. 8
   - 7

13. 9
   - 2

14. 8
   - 4

15. 8
   - 1

16. 9
   - 0

17. 10
   - 7

18. 10
   - 8

19. 4
   - 3

20. 10
   - 5

21. 10
   - 1

22. 10
   - 6

23. 8
   - 2

24. 6
   - 2

25. 6
   - 4

26. 7
   - 3

27. 4
   - 0

28. 8
   - 5

29. 7
   - 0

30. 3
   - 0

31. 5
   - 2

32. 9
   - 8

33. 3
   - 1

34. 9
   - 7

35. 5
   - 3

36. 8
   - 3

37. 10
   - 3

38. 7
   - 6

39. 6
   - 0

40. 10
   - 2

41. 8
   - 6

42. 5
   - 1

43. 9
   - 6

44. 3
   - 2

45. 10
   - 4

46. 9
   - 5

47. 9
   - 3

48. 9
   - 5

49. 6
   - 3

50. 7
   - 6

51. 6
   - 1

52. 10
   - 9

53. 2
   - 1

54. 4
   - 1

55. 3
   - 1

56. 7
   - 2

57. 5
   - 2

58. 4
   - 3

59. 7
   - 4

60. 7
   - 0

Score
/60

I Feel:

Subtracting Digits 0 - 10

Start   Finish

Date: ....   NAME : ....   Time: ....:.... ....:....

1. 5
  - 1

2. 3
  - 0

3. 4
  - 3

4. 4
  - 1

5. 5
  - 4

6. 10
  - 6

7. 9
  - 8

8. 6
  - 3

9. 8
  - 2

10. 2
  - 0

11. 7
  - 2

12. 7
  - 6

13. 4
  - 3

14. 10
  - 2

15. 5
  - 2

16. 9
  - 5

17. 8
  - 6

18. 9
  - 7

19. 8
  - 1

20. 7
  - 0

21. 9
  - 4

22. 9
  - 3

23. 9
  - 1

24. 10
  - 5

25. 8
  - 2

26. 8
  - 3

27. 7
  - 1

28. 5
  - 3

29. 9
  - 2

30. 9
  - 0

31. 4
  - 2

32. 10
  - 7

33. 9
  - 6

34. 5
  - 2

35. 6
  - 1

36. 8
  - 7

37. 3
  - 1

38. 5
  - 1

39. 6
  - 4

40. 6
  - 5

41. 7
  - 4

42. 2
  - 1

43. 8
  - 5

44. 7
  - 5

45. 10
  - 8

46. 7
  - 2

47. 6
  - 0

48. 7
  - 6

49. 4
  - 3

50. 6
  - 2

51. 7
  - 3

52. 10
  - 1

53. 10
  - 9

54. 8
  - 4

55. 9
  - 5

56. 3
  - 1

57. 10
  - 3

58. 3
  - 2

59. 10
  - 4

60. 8
  - 3

Score
/60

I Feel: 😞 😐 🙂

| 1. 4
- 3 | 2. 5
- 2 | 3. 5
- 3 | 4. 3
- 2 | 5. 6
- 5 | 6. 7
- 2 | 7. 8
- 4 | 8. 7
- 6 |

| 9. 8
- 7 | 10. 6
- 3 | 11. 6
- 2 | 12. 2
- 1 | 13. 9
- 3 | 14. 9
- 0 | 15. 8
- 2 | 16. 3
- 1 |

| 17. 10
- 6 | 18. 6
- 0 | 19. 8
- 3 | 20. 5
- 1 | 21. 9
- 4 | 22. 4
- 2 | 23. 7
- 3 | 24. 9
- 8 |

| 25. 10
- 8 | 26. 5
- 4 | 27. 9
- 7 | 28. 8
- 6 | 29. 7
- 1 | 30. 10
- 7 | 31. 7
- 0 | 32. 8
- 5 |

| 33. 6
- 4 | 34. 9
- 6 | 35. 4
- 1 | 36. 7
- 5 | 37. 10
- 5 | 38. 9
- 5 | 39. 9
- 1 | 40. 10
- 2 |

| 41. 10
- 3 | 42. 10
- 4 | 43. 8
- 1 | 44. 10
- 1 | 45. 10
- 9 | 46. 8
- 0 | 47. 3
- 2 | 48. 2
- 1 |

| 49. 9
- 7 | 50. 8
- 7 | 51. 6
- 4 | 52. 9
- 0 | 53. 6
- 3 | 54. 9
- 5 | 55. 6
- 5 | 56. 4
- 3 |

| 57. 7
- 5 | 58. 4
- 2 | 59. 6
- 5 | 60. 9
- 5 |

Score /60   I Feel: 😞 😐 🙂

1. 9 − 8

2. 10 − 2

3. 10 − 8

4. 10 − 7

5. 7 − 4

6. 2 − 1

7. 6 − 4

8. 4 − 0

9. 8 − 3

10. 7 − 1

11. 10 − 6

12. 7 − 6

13. 5 − 4

14. 6 − 2

15. 9 − 4

16. 5 − 2

17. 6 − 5

18. 6 − 5

19. 8 − 4

20. 4 − 2

21. 6 − 1

22. 10 − 4

23. 8 − 1

24. 8 − 7

25. 2 − 0

26. 9 − 2

27. 3 − 1

28. 9 − 1

29. 8 − 6

30. 5 − 1

31. 3 − 2

32. 9 − 6

33. 2 − 0

34. 5 − 0

35. 4 − 3

36. 8 − 2

37. 9 − 7

38. 10 − 9

39. 10 − 5

40. 8 − 7

41. 4 − 1

42. 9 − 5

43. 3 − 2

44. 10 − 3

45. 9 − 0

46. 7 − 5

47. 6 − 3

48. 7 − 0

49. 7 − 3

50. 5 − 4

51. 9 − 3

52. 10 − 1

53. 3 − 0

54. 6 − 0

55. 7 − 2

56. 5 − 3

57. 5 − 3

58. 8 − 5

59. 3 − 0

60. 8 − 0

Score /60

I Feel:

1.   6
    - 3

2.   5
    - 0

3.   6
    - 0

4.   9
    - 1

5.   8
    - 7

6.   9
    - 8

7.   2
    - 0

8.   5
    - 2

9.   7
    - 6

10. 9
    - 3

11. 9
    - 6

12. 6
    - 5

13. 10
    - 9

14. 10
    - 1

15. 9
    - 0

16. 5
    - 3

17. 4
    - 2

18. 10
    - 7

19. 5
    - 4

20. 9
    - 7

21. 3
    - 1

22. 4
    - 3

23. 10
    - 8

24. 2
    - 0

25. 5
    - 3

26. 8
    - 2

27. 7
    - 2

28. 10
    - 2

29. 8
    - 4

30. 8
    - 0

31. 10
    - 5

32. 5
    - 4

33. 7
    - 0

34. 9
    - 2

35. 6
    - 4

36. 2
    - 1

37. 3
    - 2

38. 6
    - 1

39. 7
    - 4

40. 3
    - 0

41. 7
    - 3

42. 7
    - 5

43. 8
    - 7

44. 9
    - 5

45. 8
    - 6

46. 8
    - 5

47. 8
    - 3

48. 3
    - 0

49. 9
    - 4

50. 6
    - 2

51. 7
    - 1

52. 10
    - 3

53. 10
    - 4

54. 10
    - 6

55. 8
    - 1

56. 4
    - 0

57. 4
    - 1

58. 6
    - 5

59. 3
    - 2

60. 5
    - 1

Score
/60

I Feel:

1. 2 − 0	2. 8 − 2	3. 8 − 1	4. 10 − 9	5. 6 − 5	6. 5 − 4	7. 10 − 2	8. 8 − 7
9. 8 − 5	10. 8 − 7	11. 3 − 1	12. 3 − 2	13. 5 − 0	14. 5 − 1	15. 3 − 0	16. 2 − 1
17. 9 − 8	18. 4 − 3	19. 6 − 5	20. 6 − 3	21. 10 − 5	22. 4 − 0	23. 7 − 4	24. 9 − 1
25. 7 − 1	26. 7 − 5	27. 8 − 6	28. 3 − 2	29. 9 − 3	30. 10 − 4	31. 5 − 4	32. 10 − 7
33. 8 − 3	34. 4 − 1	35. 3 − 0	36. 5 − 3	37. 10 − 6	38. 4 − 2	39. 8 − 0	40. 5 − 3
41. 7 − 2	42. 6 − 0	43. 10 − 1	44. 6 − 1	45. 5 − 2	46. 9 − 0	47. 6 − 2	48. 7 − 0
49. 8 − 4	50. 9 − 5	51. 9 − 6	52. 9 − 7	53. 2 − 0	54. 6 − 4	55. 7 − 3	56. 7 − 6
57. 10 − 3	58. 9 − 4	59. 10 − 8	60. 9 − 2				

Score /60

I Feel:

Date: .... NAME : .... Time: ....:.... ....:....

Start   Finish

1. 10
   - 4

2. 9
   - 0

3. 10
   - 8

4. 8
   - 4

5. 8
   - 5

6. 8
   - 2

7. 3
   - 2

8. 10
   - 9

9. 2
   - 1

10. 4
    - 0

11. 4
    - 2

12. 5
    - 2

13. 6
    - 2

14. 5
    - 3

15. 5
    - 4

16. 7
    - 5

17. 6
    - 1

18. 9
    - 4

19. 2
    - 0

20. 6
    - 0

21. 9
    - 6

22. 6
    - 5

23. 3
    - 0

24. 6
    - 3

25. 9
    - 7

26. 10
    - 5

27. 7
    - 6

28. 5
    - 1

29. 9
    - 1

30. 5
    - 3

31. 8
    - 0

32. 5
    - 4

33. 9
    - 2

34. 10
    - 2

35. 9
    - 5

36. 7
    - 2

37. 6
    - 5

38. 8
    - 6

39. 10
    - 7

40. 10
    - 3

41. 7
    - 3

42. 7
    - 0

43. 5
    - 0

44. 8
    - 7

45. 9
    - 8

46. 7
    - 1

47. 9
    - 3

48. 3
    - 0

49. 8
    - 1

50. 3
    - 1

51. 3
    - 2

52. 2
    - 0

53. 10
    - 1

54. 4
    - 1

55. 8
    - 3

56. 8
    - 7

57. 10
    - 6

58. 6
    - 4

59. 7
    - 4

60. 4
    - 3

Score /60

I Feel: ☹ 😐 🙂

1. 9   2. 9   3. 10   4. 8   5. 6   6. 10   7. 7   8. 10
  - 6   - 3   - 7   - 3   - 0   - 9   - 2   - 7

9. 8   10. 7   11. 9   12. 5   13. 5   14. 7   15. 7   16. 2
  - 6   - 0   - 7   - 0   - 3   - 6   - 4   - 0

17. 8   18. 6   19. 9   20. 5   21. 6   22. 8   23. 9   24. 4
  - 2   - 3   - 0   - 1   - 1   - 5   - 8   - 1

25. 10 26. 8   27. 6   28. 10 29. 4   30. 10 31. 9   32. 9
  - 3   - 4   - 4   - 1   - 2   - 4   - 5   - 1

33. 6   34. 5   35. 10   36. 5   37. 2   38. 7   39. 8   40. 10
  - 5   - 1   - 6   - 4   - 1   - 5   - 0   - 8

41. 8   42. 9   43. 10   44. 3   45. 8   46. 9   47. 7   48. 4
  - 1   - 4   - 2   - 2   - 7   - 2   - 1   - 3

49. 7   50. 7   51. 8   52. 3   53. 5   54. 6   55. 9   56. 10
  - 3   - 6   - 5   - 1   - 2   - 2   - 4   - 5

57. 3   58. 4   59. 7   60. 7
  - 0   - 0   - 5   - 5

Score
/60

I Feel:

Date: .... NAME : .... Time: .....:..... .....:.....

Start Finish

1. 19 - 17	2. 15 - 14	3. 19 - 12	4. 13 - 12	5. 12 - 11	6. 20 - 19	7. 10 - 10	8. 19 - 14
9. 12 - 12	10. 11 - 11	11. 19 - 16	12. 14 - 12	13. 19 - 11	14. 14 - 13	15. 18 - 13	16. 11 - 10
17. 16 - 12	18. 17 - 12	19. 13 - 11	20. 13 - 10	21. 15 - 13	22. 20 - 12	23. 18 - 12	24. 15 - 12
25. 14 - 14	26. 20 - 13	27. 16 - 14	28. 16 - 15	29. 17 - 13	30. 20 - 11	31. 16 - 16	32. 18 - 16
33. 13 - 13	34. 18 - 17	35. 15 - 15	36. 18 - 18	37. 15 - 11	38. 17 - 17	39. 17 - 15	40. 16 - 11
41. 16 - 13	42. 18 - 11	43. 14 - 11	44. 20 - 18	45. 17 - 10	46. 19 - 13	47. 14 - 10	48. 15 - 10
49. 17 - 14	50. 18 - 14	51. 17 - 11	52. 18 - 10	53. 19 - 15	54. 17 - 16	55. 20 - 17	56. 20 - 10
57. 20 - 16	58. 20 - 14	59. 19 - 18	60. 19 - 19				

Score /60

I Feel:

1.	2.	3.	4.	5.	6.	7.	8.
20	10	13	13	15	13	14	11
- 16	- 10	- 10	- 13	- 15	- 11	- 13	- 10

9.	10.	11.	12.	13.	14.	15.	16.
11	12	18	15	19	15	14	15
- 11	- 12	- 16	- 12	- 16	- 11	- 11	- 13

17.	18.	19.	20.	21.	22.	23.	24.
18	12	19	18	16	15	13	17
- 13	- 11	- 14	- 14	- 14	- 14	- 12	- 13

25.	26.	27.	28.	29.	30.	31.	32.
17	14	16	14	17	20	19	18
- 10	- 12	- 11	- 14	- 14	- 10	- 15	- 10

33.	34.	35.	36.	37.	38.	39.	40.
12	19	17	17	20	17	16	19
- 10	- 11	- 12	- 16	- 11	- 15	- 12	- 10

41.	42.	43.	44.	45.	46.	47.	48.
19	17	18	18	18	16	20	19
- 17	- 11	- 18	- 12	- 15	- 10	- 13	- 12

49.	50.	51.	52.	53.	54.	55.	56.
19	14	16	16	18	15	20	20
- 13	- 10	- 13	- 15	- 11	- 10	- 15	- 14

57.	58.	59.	60.
20	19	17	20
- 17	- 18	- 17	- 20

Score
/60

I Feel: 🙁 😐 🙂

Day
63

Date: ....

NAME : ....

Start   Finish

Time: ....:.... ....:....

| 1. | 15
- 12 | 2. | 13
- 11 | 3. | 18
- 14 | 4. | 20
- 11 | 5. | 17
- 14 | 6. | 12
- 12 | 7. | 11
- 10 | 8. | 20
- 15 |

| 9. | 20
- 13 | 10. | 13
- 12 | 11. | 14
- 11 | 12. | 16
- 15 | 13. | 17
- 11 | 14. | 16
- 13 | 15. | 10
- 10 | 16. | 15
- 13 |

| 17. | 18
- 16 | 18. | 14
- 14 | 19. | 15
- 11 | 20. | 15
- 10 | 21. | 18
- 12 | 22. | 17
- 17 | 23. | 19
- 10 | 24. | 17
- 15 |

| 25. | 13
- 10 | 26. | 18
- 13 | 27. | 14
- 13 | 28. | 18
- 11 | 29. | 20
- 17 | 30. | 15
- 14 | 31. | 19
- 14 | 32. | 14
- 10 |

| 33. | 18
- 18 | 34. | 18
- 10 | 35. | 17
- 10 | 36. | 20
- 16 | 37. | 18
- 15 | 38. | 11
- 11 | 39. | 19
- 16 | 40. | 19
- 18 |

| 41. | 19
- 17 | 42. | 16
- 11 | 43. | 20
- 20 | 44. | 20
- 14 | 45. | 17
- 13 | 46. | 14
- 12 | 47. | 17
- 16 | 48. | 13
- 13 |

| 49. | 19
- 15 | 50. | 16
- 14 | 51. | 20
- 10 | 52. | 19
- 13 | 53. | 12
- 11 | 54. | 15
- 15 | 55. | 19
- 11 | 56. | 17
- 12 |

| 57. | 16
- 12 | 58. | 19
- 12 | 59. | 16
- 10 | 60. | 12
- 10 |

Score
/60

I Feel: ☹ 😐 🙂

1. 20
   - 14

2. 16
   - 10

3. 14
   - 10

4. 12
   - 10

5. 19
   - 12

6. 17
   - 17

7. 20
   - 10

8. 18
   - 15

9. 10
   - 10

10. 19
    - 18

11. 15
    - 10

12. 17
    - 11

13. 14
    - 12

14. 19
    - 13

15. 14
    - 13

16. 18
    - 16

17. 20
    - 15

18. 17
    - 15

19. 13
    - 12

20. 18
    - 12

21. 17
    - 13

22. 11
    - 11

23. 19
    - 14

24. 17
    - 10

25. 16
    - 11

26. 20
    - 16

27. 12
    - 12

28. 19
    - 16

29. 15
    - 15

30. 12
    - 11

31. 18
    - 18

32. 17
    - 12

33. 19
    - 10

34. 18
    - 14

35. 14
    - 11

36. 20
    - 17

37. 16
    - 12

38. 13
    - 13

39. 17
    - 16

40. 20
    - 20

41. 20
    - 11

42. 15
    - 12

43. 18
    - 13

44. 15
    - 14

45. 15
    - 13

46. 13
    - 10

47. 14
    - 14

48. 16
    - 15

49. 13
    - 11

50. 18
    - 10

51. 19
    - 17

52. 16
    - 14

53. 19
    - 15

54. 17
    - 14

55. 19
    - 11

56. 15
    - 11

57. 16
    - 13

58. 18
    - 11

59. 11
    - 10

60. 20
    - 13

Score
/60

I Feel: 😊

Subtracting Digits 10-20

Start   Finish

Date: ....   NAME : ....   Time: ....:..... ....:.....

1. 14 - 13	2. 19 - 14	3. 20 - 13	4. 19 - 11	5. 16 - 10	6. 15 - 12	7. 14 - 14	8. 17 - 14
9. 18 - 11	10. 19 - 17	11. 18 - 13	12. 19 - 15	13. 16 - 15	14. 18 - 16	15. 16 - 14	16. 18 - 15
17. 16 - 13	18. 12 - 10	19. 20 - 10	20. 19 - 12	21. 16 - 12	22. 15 - 11	23. 11 - 11	24. 13 - 12
25. 17 - 11	26. 14 - 12	27. 20 - 16	28. 14 - 10	29. 15 - 13	30. 20 - 20	31. 20 - 11	32. 20 - 17
33. 15 - 14	34. 19 - 16	35. 13 - 13	36. 17 - 16	37. 18 - 10	38. 13 - 11	39. 18 - 12	40. 12 - 12
41. 17 - 13	42. 17 - 10	43. 18 - 14	44. 13 - 10	45. 17 - 12	46. 17 - 17	47. 18 - 18	48. 10 - 10
49. 16 - 11	50. 19 - 13	51. 11 - 10	52. 19 - 18	53. 14 - 11	54. 15 - 10	55. 20 - 15	56. 17 - 15
57. 19 - 10	58. 12 - 11	59. 15 - 15	60. 20 - 14				

Score /60

I Feel:

Subtracting Digits 10-20

Date: ....   NAME : ....   Start   Finish

Time: ....:....   ....:....

1. 12 - 10	2. 17 - 12	3. 16 - 10	4. 19 - 15	5. 19 - 11	6. 19 - 14	7. 18 - 11	8. 14 - 12
9. 19 - 16	10. 19 - 17	11. 16 - 13	12. 15 - 12	13. 18 - 14	14. 18 - 12	15. 17 - 13	16. 16 - 12
17. 16 - 14	18. 15 - 14	19. 17 - 16	20. 14 - 14	21. 20 - 20	22. 17 - 11	23. 18 - 10	24. 13 - 11
25. 17 - 15	26. 18 - 18	27. 15 - 10	28. 17 - 17	29. 15 - 11	30. 20 - 17	31. 14 - 11	32. 20 - 10
33. 16 - 15	34. 12 - 12	35. 19 - 12	36. 19 - 13	37. 20 - 16	38. 13 - 10	39. 19 - 10	40. 19 - 18
41. 14 - 13	42. 14 - 10	43. 17 - 10	44. 15 - 13	45. 18 - 13	46. 12 - 11	47. 13 - 13	48. 11 - 11
49. 10 - 10	50. 20 - 14	51. 17 - 14	52. 20 - 11	53. 18 - 16	54. 11 - 10	55. 13 - 12	56. 15 - 15
57. 20 - 13	58. 16 - 11	59. 18 - 15	60. 20 - 15				

Score /60

I Feel:

Subtracting Digits 10-20

Start   Finish

Date: ....   NAME : ....   Time: .....:.....   .....:.....

1.	2.	3.	4.	5.	6.	7.	8.
19 − 15	19 − 10	13 − 10	20 − 17	13 − 11	18 − 14	19 − 12	17 − 16

9.	10.	11.	12.	13.	14.	15.	16.
19 − 11	17 − 13	19 − 16	18 − 18	20 − 10	16 − 13	18 − 11	15 − 10

17.	18.	19.	20.	21.	22.	23.	24.
20 − 11	17 − 10	16 − 12	20 − 20	20 − 16	14 − 10	19 − 13	16 − 10

25.	26.	27.	28.	29.	30.	31.	32.
14 − 11	16 − 11	15 − 14	18 − 13	10 − 10	14 − 13	18 − 16	11 − 10

33.	34.	35.	36.	37.	38.	39.	40.
11 − 11	14 − 14	15 − 15	16 − 15	15 − 11	19 − 18	20 − 14	17 − 17

41.	42.	43.	44.	45.	46.	47.	48.
13 − 12	17 − 12	18 − 12	12 − 10	12 − 11	17 − 15	20 − 15	20 − 13

49.	50.	51.	52.	53.	54.	55.	56.
18 − 15	18 − 10	13 − 13	17 − 11	15 − 13	17 − 14	15 − 12	12 − 12

57.	58.	59.	60.
19 − 17	16 − 14	14 − 12	19 − 14

Score /60

I Feel:

1. 20
    - 12

2. 10
    - 10

3. 14
    - 11

4. 17
    - 12

5. 14
    - 13

6. 17
    - 13

7. 13
    - 11

8. 16
    - 13

9. 11
    - 11

10. 19
    - 14

11. 11
    - 10

12. 18
    - 15

13. 14
    - 12

14. 17
    - 15

15. 17
    - 17

16. 19
    - 15

17. 13
    - 12

18. 12
    - 11

19. 17
    - 10

20. 19
    - 16

21. 12
    - 10

22. 18
    - 16

23. 18
    - 12

24. 12
    - 12

25. 19
    - 11

26. 16
    - 14

27. 20
    - 15

28. 14
    - 10

29. 16
    - 11

30. 20
    - 14

31. 15
    - 12

32. 13
    - 10

33. 18
    - 18

34. 19
    - 10

35. 19
    - 17

36. 18
    - 13

37. 13
    - 13

38. 18
    - 10

39. 15
    - 10

40. 16
    - 12

41. 17
    - 14

42. 16
    - 10

43. 20
    - 17

44. 15
    - 15

45. 19
    - 18

46. 16
    - 16

47. 15
    - 13

48. 19
    - 13

49. 15
    - 14

50. 18
    - 17

51. 17
    - 16

52. 15
    - 11

53. 20
    - 13

54. 17
    - 11

55. 16
    - 15

56. 19
    - 12

57. 18
    - 14

58. 14
    - 14

59. 20
    - 16

60. 20
    - 18

Score /60

I Feel:

1. 20
  - 4

2. 11
  - 9

3. 9
  - 6

4. 17
  - 10

5. 11
  - 5

6. 19
  - 11

7. 14
  - 4

8. 19
  - 10

9. 11
  - 10

10. 9
  - 4

11. 11
  - 6

12. 2
  - 2

13. 18
  - 10

14. 17
  - 12

15. 11
  - 3

16. 11
  - 11

17. 19
  - 12

18. 0
  - 0

19. 13
  - 13

20. 4
  - 0

21. 10
  - 6

22. 19
  - 14

23. 8
  - 0

24. 15
  - 9

25. 18
  - 6

26. 4
  - 3

27. 15
  - 3

28. 5
  - 1

29. 2
  - 1

30. 1
  - 1

31. 11
  - 2

32. 14
  - 5

33. 18
  - 11

34. 10
  - 2

35. 13
  - 2

36. 5
  - 2

37. 10
  - 8

38. 13
  - 8

39. 19
  - 5

40. 20
  - 15

41. 20
  - 13

42. 14
  - 8

43. 7
  - 0

44. 18
  - 14

45. 18
  - 3

46. 14
  - 9

47. 14
  - 13

48. 9
  - 2

49. 4
  - 2

50. 16
  - 2

51. 19
  - 9

52. 11
  - 7

53. 5
  - 3

54. 14
  - 0

55. 15
  - 4

56. 14
  - 11

57. 8
  - 5

58. 17
  - 2

59. 10
  - 10

60. 12
  - 8

Score
/60

I Feel: :(:| :)

1. 13
    - 2

2. 5
    - 2

3. 14
    - 9

4. 19
    - 11

5. 20
    - 13

6. 18
    - 6

7. 19
    - 10

8. 17
    - 10

9. 4
    - 0

10. 5
    - 1

11. 10
    - 6

12. 18
    - 14

13. 10
    - 2

14. 13
    - 13

15. 5
    - 3

16. 0
    - 0

17. 16
    - 2

18. 14
    - 4

19. 9
    - 6

20. 13
    - 8

21. 14
    - 5

22. 20
    - 15

23. 2
    - 1

24. 11
    - 10

25. 2
    - 2

26. 8
    - 0

27. 8
    - 5

28. 18
    - 3

29. 15
    - 4

30. 7
    - 0

31. 11
    - 5

32. 1
    - 1

33. 14
    - 0

34. 4
    - 2

35. 11
    - 7

36. 17
    - 2

37. 12
    - 8

38. 19
    - 14

39. 14
    - 11

40. 10
    - 10

41. 14
    - 8

42. 11
    - 2

43. 19
    - 5

44. 15
    - 3

45. 10
    - 8

46. 17
    - 12

47. 15
    - 9

48. 11
    - 6

49. 11
    - 9

50. 20
    - 4

51. 9
    - 4

52. 14
    - 13

53. 18
    - 11

54. 11
    - 11

55. 11
    - 3

56. 19
    - 9

57. 18
    - 10

58. 4
    - 3

59. 9
    - 2

60. 19
    - 12

Score /60

I Feel: ☹ 😐 🙂

 Start   Finish

Date: ....   NAME : ....   Time: ....:.... ....:.....

1. 14 − 0	2. 18 − 3	3. 10 − 2	4. 14 − 8	5. 14 − 5	6. 11 − 11	7. 9 − 4	8. 11 − 6
9. 7 − 0	10. 4 − 3	11. 14 − 4	12. 2 − 1	13. 5 − 1	14. 19 − 11	15. 17 − 12	16. 14 − 13
17. 10 − 8	18. 15 − 3	19. 16 − 2	20. 11 − 2	21. 19 − 5	22. 17 − 2	23. 8 − 5	24. 15 − 9
25. 4 − 2	26. 5 − 3	27. 14 − 11	28. 13 − 8	29. 11 − 10	30. 18 − 11	31. 10 − 10	32. 18 − 14
33. 20 − 15	34. 11 − 5	35. 9 − 6	36. 10 − 6	37. 2 − 2	38. 20 − 13	39. 4 − 0	40. 18 − 10
41. 1 − 1	42. 20 − 4	43. 19 − 14	44. 18 − 6	45. 11 − 9	46. 11 − 7	47. 15 − 4	48. 12 − 8
49. 8 − 0	50. 13 − 13	51. 5 − 2	52. 17 − 10	53. 19 − 10	54. 14 − 9	55. 19 − 9	56. 19 − 12
57. 9 − 2	58. 0 − 0	59. 13 − 2	60. 11 − 3				

Score /60   I Feel:

Day 72

Subtracting Digits 0-20

Date: ....   NAME : ....   Start   Finish

Time: ....:.... ....:....

1. 19 - 16	2. 5 - 2	3. 18 - 3	4. 17 - 5	5. 12 - 3	6. 14 - 9	7. 20 - 18	8. 12 - 6
9. 2 - 0	10. 11 - 10	11. 11 - 7	12. 10 - 1	13. 15 - 13	14. 3 - 3	15. 20 - 8	16. 8 - 4
17. 6 - 3	18. 9 - 4	19. 7 - 7	20. 13 - 12	21. 11 - 1	22. 16 - 9	23. 11 - 2	24. 8 - 3
25. 4 - 2	26. 13 - 0	27. 8 - 8	28. 3 - 1	29. 16 - 2	30. 11 - 4	31. 3 - 0	32. 18 - 16
33. 1 - 0	34. 2 - 1	35. 16 - 5	36. 0 - 0	37. 17 - 7	38. 5 - 5	39. 17 - 10	40. 19 - 6
41. 17 - 12	42. 3 - 2	43. 15 - 7	44. 12 - 9	45. 11 - 9	46. 2 - 2	47. 12 - 12	48. 5 - 4
49. 4 - 1	50. 10 - 3	51. 10 - 5	52. 6 - 2	53. 6 - 1	54. 10 - 4	55. 9 - 2	56. 17 - 8
57. 18 - 1	58. 11 - 3	59. 11 - 0	60. 18 - 2				

Score /60

I Feel: :(:| :)

Subtracting Digits 0-20

Start   Finish

Date: ....   NAME : ....   Time: ....:.... ....:....

1.	2.	3.	4.	5.	6.	7.	8.
12	14	8	6	17	11	3	11
- 3	- 9	- 3	- 1	- 8	- 10	- 1	- 7

9.	10.	11.	12.	13.	14.	15.	16.
9	18	4	8	11	7	16	5
- 4	- 1	- 2	- 4	- 4	- 7	- 5	- 4

17.	18.	19.	20.	21.	22.	23.	24.
3	5	1	4	16	12	11	12
- 0	- 5	- 0	- 1	- 9	- 12	- 3	- 6

25.	26.	27.	28.	29.	30.	31.	32.
20	17	9	11	13	17	16	6
- 18	- 10	- 2	- 1	- 12	- 7	- 2	- 2

33.	34.	35.	36.	37.	38.	39.	40.
17	17	15	15	18	19	11	10
- 12	- 5	- 7	- 13	- 3	- 6	- 9	- 3

41.	42.	43.	44.	45.	46.	47.	48.
2	18	2	20	10	10	12	2
- 1	- 16	- 0	- 8	- 1	- 5	- 9	- 2

49.	50.	51.	52.	53.	54.	55.	56.
8	3	11	18	3	19	10	6
- 8	- 2	- 0	- 2	- 3	- 16	- 4	- 3

57.	58.	59.	60.
5	11	0	13
- 2	- 2	- 0	- 0

Score /60

I Feel:

1. 8 − 4	2. 11 − 10	3. 8 − 8	4. 18 − 16	5. 13 − 12	6. 17 − 7	7. 19 − 6	8. 3 − 0
9. 15 − 13	10. 5 − 4	11. 2 − 0	12. 19 − 16	13. 10 − 4	14. 2 − 1	15. 4 − 2	16. 11 − 0
17. 11 − 9	18. 18 − 3	19. 13 − 0	20. 16 − 9	21. 3 − 3	22. 11 − 7	23. 12 − 9	24. 15 − 7
25. 17 − 5	26. 12 − 6	27. 17 − 10	28. 11 − 1	29. 11 − 4	30. 4 − 1	31. 8 − 3	32. 6 − 3
33. 3 − 2	34. 5 − 5	35. 6 − 2	36. 20 − 18	37. 10 − 1	38. 1 − 0	39. 2 − 2	40. 10 − 3
41. 6 − 1	42. 16 − 5	43. 3 − 1	44. 18 − 1	45. 0 − 0	46. 12 − 3	47. 17 − 8	48. 20 − 8
49. 5 − 2	50. 14 − 9	51. 16 − 2	52. 9 − 2	53. 7 − 7	54. 11 − 3	55. 17 − 12	56. 9 − 4
57. 10 − 5	58. 12 − 12	59. 18 − 2	60. 11 − 2				

Score /60

I Feel:

1. 15 - 7	2. 5 - 4	3. 10 - 5	4. 3 - 1	5. 16 - 5	6. 12 - 12	7. 11 - 4	8. 18 - 16
9. 11 - 3	10. 6 - 3	11. 11 - 0	12. 5 - 5	13. 7 - 7	14. 8 - 3	15. 19 - 6	16. 0 - 0
17. 17 - 7	18. 4 - 1	19. 17 - 8	20. 2 - 1	21. 11 - 9	22. 10 - 4	23. 17 - 5	24. 11 - 1
25. 6 - 2	26. 2 - 2	27. 18 - 3	28. 3 - 3	29. 16 - 2	30. 8 - 8	31. 20 - 8	32. 8 - 4
33. 4 - 2	34. 11 - 10	35. 11 - 2	36. 10 - 1	37. 2 - 0	38. 17 - 12	39. 14 - 9	40. 1 - 0
41. 12 - 6	42. 3 - 0	43. 12 - 3	44. 9 - 4	45. 17 - 10	46. 11 - 7	47. 5 - 2	48. 12 - 9
49. 13 - 12	50. 6 - 1	51. 10 - 3	52. 16 - 9	53. 9 - 2	54. 18 - 2	55. 13 - 0	56. 20 - 18
57. 3 - 2	58. 15 - 13	59. 19 - 16	60. 18 - 1				

Score /60

I Feel: ☹ 😐 🙂

Day 76

Date: ....   NAME : ....   Start   Finish

Time: ....:.... ....:....

1. 20
    - 8

2. 17
    - 12

3. 0
    - 0

4. 11
    - 2

5. 10
    - 5

6. 4
    - 2

7. 11
    - 4

8. 2
    - 2

9. 8
    - 4

10. 10
     - 4

11. 13
     - 12

12. 5
     - 5

13. 17
     - 10

14. 11
     - 10

15. 10
     - 3

16. 16
     - 9

17. 16
     - 2

18. 13
     - 0

19. 16
     - 5

20. 11
     - 3

21. 11
     - 9

22. 12
     - 12

23. 18
     - 2

24. 11
     - 1

25. 20
     - 18

26. 17
     - 7

27. 8
     - 8

28. 3
     - 0

29. 12
     - 6

30. 3
     - 1

31. 6
     - 3

32. 9
     - 2

33. 18
     - 1

34. 2
     - 1

35. 15
     - 13

36. 18
     - 16

37. 5
     - 4

38. 7
     - 7

39. 3
     - 3

40. 1
     - 0

41. 2
     - 0

42. 17
     - 8

43. 6
     - 2

44. 14
     - 9

45. 9
     - 4

46. 5
     - 2

47. 8
     - 3

48. 4
     - 1

49. 12
     - 9

50. 11
     - 0

51. 11
     - 7

52. 12
     - 3

53. 6
     - 1

54. 15
     - 7

55. 19
     - 16

56. 18
     - 3

57. 3
     - 2

58. 10
     - 1

59. 17
     - 5

60. 19
     - 6

Score
/60

I Feel: :(:| :)

1.	2.	3.	4.	5.	6.	7.	8.
12	9	8	6	0	19	6	20
- 8	- 4	- 2	- 4	- 0	- 2	- 3	- 10

9.	10.	11.	12.	13.	14.	15.	16.
2	17	17	13	18	6	11	16
- 1	- 10	- 6	- 3	- 8	- 6	- 3	- 8

17.	18.	19.	20.	21.	22.	23.	24.
15	2	10	15	10	10	15	16
- 9	- 2	- 0	- 11	- 5	- 9	- 8	- 3

25.	26.	27.	28.	29.	30.	31.	32.
4	10	3	12	11	19	16	3
- 3	- 6	- 1	- 1	- 2	- 6	- 1	- 3

33.	34.	35.	36.	37.	38.	39.	40.
6	5	5	16	1	16	7	7
- 1	- 3	- 2	- 11	- 0	- 2	- 3	- 2

41.	42.	43.	44.	45.	46.	47.	48.
10	19	12	14	18	7	14	15
- 4	- 5	- 9	- 5	- 6	- 4	- 14	- 10

49.	50.	51.	52.	53.	54.	55.	56.
18	13	20	8	9	1	18	8
- 14	- 6	- 11	- 1	- 5	- 1	- 5	- 7

57.	58.	59.	60.
16	12	7	19
- 9	- 4	- 1	- 14

Score /60

I Feel: 😊

1. 8 - 2	2. 12 - 1	3. 7 - 1	4. 6 - 6	5. 2 - 2	6. 10 - 0	7. 10 - 9	8. 6 - 1
9. 12 - 9	10. 12 - 4	11. 19 - 5	12. 19 - 2	13. 11 - 2	14. 15 - 10	15. 16 - 1	16. 18 - 6
17. 16 - 3	18. 20 - 10	19. 11 - 3	20. 9 - 5	21. 6 - 4	22. 14 - 14	23. 13 - 6	24. 10 - 6
25. 5 - 3	26. 17 - 6	27. 8 - 1	28. 3 - 1	29. 12 - 8	30. 18 - 14	31. 4 - 3	32. 8 - 7
33. 10 - 5	34. 5 - 2	35. 3 - 3	36. 1 - 0	37. 16 - 2	38. 16 - 9	39. 19 - 6	40. 18 - 8
41. 17 - 10	42. 15 - 11	43. 16 - 8	44. 15 - 9	45. 18 - 5	46. 13 - 3	47. 9 - 4	48. 19 - 14
49. 16 - 11	50. 15 - 8	51. 10 - 4	52. 2 - 1	53. 20 - 11	54. 0 - 0	55. 7 - 3	56. 7 - 2
57. 7 - 4	58. 14 - 5	59. 1 - 1	60. 6 - 3				

Score
/60

I Feel:

Date: ....   NAME : ....   Time: ....:.... ....:....

Start   Finish

1. 16   2. 10   3. 12   4. 15   5. 0   6. 18   7. 14   8. 7
  - 9   - 0   - 4   - 11   - 0   - 6   - 14   - 3

9. 20   10. 3   11. 10   12. 17   13. 9   14. 6   15. 7   16. 13
  - 11   - 1   - 5   - 6   - 4   - 6   - 1   - 3

17. 19   18. 15   19. 18   20. 5   21. 2   22. 8   23. 6   24. 16
  - 6   - 8   - 14   - 3   - 1   - 1   - 4   - 2

25. 10   26. 12   27. 12   28. 16   29. 16   30. 9   31. 18   32. 3
  - 9   - 9   - 8   - 11   - 3   - 5   - 5   - 3

33. 8   34. 11   35. 4   36. 19   37. 12   38. 16   39. 15   40. 8
  - 2   - 2   - 3   - 5   - 1   - 8   - 9   - 7

41. 19   42. 10   43. 5   44. 6   45. 18   46. 6   47. 20   48. 13
  - 14   - 6   - 2   - 1   - 8   - 3   - 10   - 6

49. 16   50. 17   51. 10   52. 14   53. 7   54. 7   55. 11   56. 19
  - 1   - 10   - 4   - 5   - 4   - 2   - 3   - 2

57. 2   58. 15   59. 1   60. 1
  - 2   - 10   - 1   - 0

Score /60

I Feel: 😞 😐 🙂

Subtracting Digits 0-20

Date: ....   NAME : ....

Start   Finish

Time: ....:....   ....:....

1. 17 − 4	2. 6 − 5	3. 13 − 12	4. 15 − 3	5. 7 − 6	6. 17 − 9	7. 16 − 7	8. 6 − 4
9. 12 − 8	10. 14 − 9	11. 17 − 7	12. 4 − 1	13. 20 − 18	14. 8 − 4	15. 11 − 10	16. 10 − 10
17. 16 − 0	18. 18 − 10	19. 8 − 5	20. 0 − 0	21. 8 − 8	22. 5 − 2	23. 13 − 7	24. 6 − 2
25. 10 − 4	26. 11 − 5	27. 9 − 6	28. 18 − 5	29. 17 − 17	30. 19 − 5	31. 15 − 9	32. 5 − 3
33. 19 − 13	34. 2 − 1	35. 3 − 2	36. 14 − 10	37. 19 − 12	38. 12 − 2	39. 12 − 7	40. 10 − 2
41. 17 − 5	42. 4 − 0	43. 12 − 0	44. 1 − 0	45. 10 − 3	46. 11 − 1	47. 19 − 10	48. 14 − 6
49. 5 − 0	50. 3 − 3	51. 14 − 13	52. 20 − 3	53. 18 − 6	54. 5 − 1	55. 13 − 1	56. 5 − 4
57. 1 − 1	58. 16 − 10	59. 6 − 6	60. 18 − 7				

Score /60

I Feel: ☹ 😐 🙂

1. 2
    + 9

2. 2
    + 12

3. 12
    - 3

4. 17
    - 6

5. 8
    + 15

6. 9
    - 8

7. 14
    - 8

8. 19
    + 3

9. 4
    - 2

10. 11
    + 19

11. 1
    - 1

12. 15
    - 12

13. 15
    + 19

14. 5
    + 18

15. 2
    - 0

16. 5
    + 10

17. 8
    - 4

18. 9
    + 5

19. 16
    - 11

20. 13
    - 2

21. 19
    - 16

22. 2
    - 2

23. 14
    + 15

24. 16
    - 4

25. 19
    + 13

26. 19
    - 6

27. 5
    - 2

28. 18
    - 3

29. 14
    + 18

30. 6
    + 8

31. 2
    + 15

32. 6
    + 3

33. 9
    - 5

34. 3
    - 3

35. 9
    + 17

36. 2
    - 1

37. 5
    + 13

38. 14
    + 13

39. 9
    + 9

40. 17
    - 16

41. 8
    - 5

42. 8
    - 7

43. 10
    + 18

44. 15
    - 11

45. 13
    + 15

46. 11
    + 18

47. 10
    + 17

48. 11
    - 8

49. 12
    + 8

50. 1
    + 3

51. 9
    - 6

52. 4
    + 12

53. 17
    - 15

54. 16
    - 16

55. 13
    - 10

56. 15
    + 9

57. 17
    + 9

58. 15
    + 3

59. 15
    - 4

60. 5
    + 20

Score
/60

I Feel:

1. 3
    - 3

2. 9
    + 9

3. 8
    - 4

4. 17
    - 15

5. 14
    + 15

6. 1
    + 3

7. 13
    - 2

8. 9
    - 6

9. 14
    + 18

10. 1
    - 1

11. 2
    + 12

12. 10
    + 18

13. 2
    + 9

14. 4
    - 2

15. 5
    - 2

16. 14
    - 8

17. 12
    + 8

18. 17
    - 6

19. 9
    - 5

20. 6
    + 8

21. 19
    - 16

22. 2
    - 2

23. 17
    + 9

24. 15
    + 19

25. 10
    + 17

26. 4
    + 12

27. 15
    - 11

28. 5
    + 10

29. 15
    - 12

30. 2
    - 0

31. 16
    - 16

32. 18
    - 3

33. 2
    - 1

34. 14
    + 13

35. 19
    - 6

36. 11
    + 18

37. 19
    + 13

38. 5
    + 13

39. 9
    + 17

40. 15
    - 4

41. 16
    - 4

42. 9
    + 5

43. 2
    + 15

44. 5
    + 18

45. 11
    - 8

46. 13
    - 10

47. 8
    - 5

48. 13
    + 15

49. 16
    - 11

50. 8
    - 7

51. 8
    + 15

52. 17
    - 16

53. 9
    - 8

54. 11
    + 19

55. 5
    + 20

56. 19
    + 3

57. 15
    + 9

58. 12
    - 3

59. 6
    + 3

60. 15
    + 3

Score /60

I Feel:

1.	2.	3.	4.	5.	6.	7.	8.
14	10	11	12	2	6	2	0
- 9	+ 0	- 2	+ 7	+ 19	+ 14	- 1	+ 3

9.	10.	11.	12.	13.	14.	15.	16.
10	0	18	2	3	6	11	14
+ 20	+ 1	- 14	+ 12	- 1	- 3	- 10	+ 8

17.	18.	19.	20.	21.	22.	23.	24.
6	14	5	14	9	2	11	17
- 5	- 7	- 3	+ 17	+ 3	+ 11	+ 11	- 14

25.	26.	27.	28.	29.	30.	31.	32.
13	4	14	15	16	17	13	18
+ 13	- 2	- 13	- 4	+ 11	+ 11	- 8	- 17

33.	34.	35.	36.	37.	38.	39.	40.
1	13	13	12	19	17	17	15
+ 5	- 2	- 6	+ 9	- 4	+ 16	- 12	+ 14

41.	42.	43.	44.	45.	46.	47.	48.
15	1	15	8	13	14	8	19
+ 1	+ 9	- 3	+ 9	- 11	+ 7	- 3	+ 13

49.	50.	51.	52.	53.	54.	55.	56.
19	3	20	15	12	17	18	11
- 17	- 2	- 11	- 12	+ 3	- 9	+ 10	+ 15

57.	58.	59.	60.
1	6	0	19
+ 11	- 0	- 0	+ 16

Score
/60

I Feel: 😞 😐 🙂

Mixed Problems

 Start   Finish

Date: ....   NAME : ....   Time: ....:.... ....:....

1. 2
   + 12

2. 11
   - 2

3. 14
   + 7

4. 11
   - 10

5. 6
   - 3

6. 9
   + 3

7. 17
   - 14

8. 18
   + 10

9. 17
   - 12

10. 6
   - 5

11. 18
   - 17

12. 19
   - 4

13. 14
   - 9

14. 15
   - 3

15. 19
   + 16

16. 1
   + 11

17. 12
   + 3

18. 8
   + 9

19. 12
   + 9

20. 2
   - 1

21. 6
   - 0

22. 1
   + 9

23. 14
   - 13

24. 0
   + 1

25. 3
   - 2

26. 11
   + 11

27. 13
   + 13

28. 12
   + 7

29. 13
   - 11

30. 19
   - 17

31. 0
   + 3

32. 8
   - 3

33. 15
   - 12

34. 17
   + 11

35. 5
   - 3

36. 20
   - 11

37. 4
   - 2

38. 16
   + 11

39. 14
   + 8

40. 18
   - 14

41. 13
   - 2

42. 13
   - 8

43. 14
   - 7

44. 14
   + 17

45. 17
   - 9

46. 11
   + 15

47. 0
   - 0

48. 1
   + 5

49. 10
   + 0

50. 2
   + 11

51. 15
   + 1

52. 3
   - 1

53. 2
   + 19

54. 13
   - 6

55. 19
   + 13

56. 17
   + 16

57. 15
   - 4

58. 15
   + 14

59. 10
   + 20

60. 6
   + 14

Score /60

I Feel: 😞 😐 🙂

1.   4
   + 18

2. 14
  - 10

3. 11
  - 4

4. 18
  + 19

5.   5
   - 3

6. 14
  - 11

7. 13
  - 11

8.   8
   - 3

9.   7
   + 1

10. 12
   - 11

11. 14
   - 5

12. 17
   - 1

13. 17
   + 11

14. 20
   + 4

15.   9
    - 2

16.   6
    + 9

17. 16
   + 14

18.   7
    - 6

19.   1
    + 6

20.   5
    - 2

21. 11
   + 14

22.   7
    + 7

23. 15
    - 4

24.   6
   + 11

25.   8
   + 9

26. 13
   + 6

27.   6
    + 2

28.   1
   + 12

29. 19
   + 13

30. 16
   + 19

31.   9
    - 3

32.   6
    - 6

33.   1
   + 4

34. 14
   + 14

35.   2
    - 1

36.   5
    - 1

37. 12
    - 8

38.   7
   + 19

39.   2
    - 2

40. 13
   + 12

41. 19
   + 9

42. 16
   + 5

43.   5
    - 0

44.   9
    + 5

45. 16
   - 15

46.   4
    - 3

47.   7
    - 2

48. 14
   + 17

49.   7
   + 8

50.   8
    - 6

51.   7
   + 18

52. 12
    - 3

53.   4
    - 2

54.   9
    - 1

55. 14
    - 0

56. 15
    - 9

57. 15
   + 13

58. 10
   + 9

59.   6
    + 1

60. 17
    - 4

Score
/60

I Feel: 😞 😐 🙂

1. 7
   + 3

2. 19
   + 1

3. 8
   - 3

4. 12
   - 7

5. 7
   + 20

6. 16
   - 5

7. 4
   - 1

8. 10
   + 18

9. 5
   - 2

10. 13
   + 12

11. 20
   - 16

12. 14
   - 13

13. 9
   - 9

14. 9
   + 4

15. 6
   + 1

16. 17
   + 0

17. 6
   + 16

18. 8
   + 11

19. 16
   + 7

20. 3
   - 1

21. 16
   - 11

22. 13
   + 1

23. 5
   - 3

24. 2
   - 2

25. 9
   - 6

26. 11
   - 9

27. 3
   + 15

28. 1
   + 5

29. 10
   + 5

30. 1
   - 1

31. 0
   - 0

32. 13
   - 8

33. 0
   + 17

34. 9
   + 1

35. 9
   - 5

36. 2
   - 1

37. 4
   - 3

38. 12
   - 10

39. 17
   + 12

40. 4
   - 2

41. 16
   + 18

42. 16
   + 12

43. 17
   + 16

44. 18
   + 14

45. 18
   + 17

46. 11
   - 6

47. 15
   - 9

48. 16
   - 4

49. 20
   + 8

50. 20
   - 14

51. 20
   + 17

52. 6
   - 5

53. 4
   + 4

54. 8
   + 16

55. 3
   - 0

56. 10
   + 0

57. 15
   - 6

58. 18
   + 12

59. 7
   - 5

60. 18
   + 5

Score
/60

I Feel: ☹ 😐 🙂

1.   7
    - 4

2. 15
   - 4

3. 14
   - 6

4.   6
  + 11

5. 14
  + 19

6.   7
   - 3

7.   8
   - 2

8.   5
   - 4

9. 13
  + 13

10. 19
  + 20

11. 9
   - 2

12. 5
  + 17

13. 18
  + 5

14. 2
  + 11

15. 14
   - 8

16. 20
  + 19

17. 9
  + 17

18. 6
  + 13

19. 18
   - 3

20. 13
  + 15

21. 11
   - 7

22. 5
  + 3

23. 14
  + 10

24. 8
  + 7

25. 9
  + 19

26. 10
   - 5

27. 1
   - 1

28. 7
  + 7

29. 7
   - 7

30. 13
  + 2

31. 5
   - 5

32. 8
   - 1

33. 9
   - 4

34. 14
   - 4

35. 8
  + 1

36. 7
   - 6

37. 12
  + 8

38. 9
  + 18

39. 17
  - 14

40. 8
   - 6

41. 3
   - 2

42. 14
  + 3

43. 12
  + 10

44. 13
  + 11

45. 11
   - 6

46. 3
  + 7

47. 10
  - 10

48. 9
   - 3

49. 12
   - 2

50. 12
   - 5

51. 12
   - 1

52. 16
  + 6

53. 5
  + 1

54. 15
  + 11

55. 15
  + 16

56. 16
  + 13

57. 19
   - 1

58. 8
  + 14

59. 2
   - 2

60. 6
   - 6

Score /60

I Feel: ☹ ☺ ☺

1. 9
   - 2

2. 9
   - 4

3. 20
   + 19

4. 3
   - 2

5. 19
   - 1

6. 5
   + 17

7. 9
   + 19

8. 19
   + 20

9. 2
   + 11

10. 12
    + 8

11. 5
    + 3

12. 7
    - 3

13. 16
    + 13

14. 14
    + 19

15. 14
    - 4

16. 8
    + 14

17. 13
    + 15

18. 8
    + 7

19. 7
    - 4

20. 13
    + 2

21. 13
    + 11

22. 7
    - 7

23. 6
    + 11

24. 13
    + 13

25. 15
    + 11

26. 15
    + 16

27. 6
    - 6

28. 14
    + 10

29. 12
    - 5

30. 14
    + 3

31. 9
    + 18

32. 5
    + 1

33. 14
    - 6

34. 8
    - 2

35. 7
    + 7

36. 10
    - 10

37. 18
    - 3

38. 17
    - 14

39. 2
    - 2

40. 5
    - 4

41. 12
    - 1

42. 7
    - 6

43. 10
    - 5

44. 11
    - 6

45. 18
    + 5

46. 16
    + 6

47. 12
    + 10

48. 1
    - 1

49. 8
    - 6

50. 14
    - 8

51. 9
    + 17

52. 6
    + 13

53. 12
    - 2

54. 8
    + 1

55. 9
    - 3

56. 8
    - 1

57. 11
    - 7

58. 15
    - 4

59. 3
    + 7

60. 5
    - 5

Score /60

I Feel: 😟 😐 🙂

1. 14   2. 3   3. 13   4. 14   5. 5   6. 15   7. 20   8. 6
    + 19   - 2   + 11   - 6   - 5   + 16   + 19   + 13

9. 10   10. 18   11. 5   12. 11   13. 7   14. 11   15. 9   16. 5
    - 5   + 5   - 4   - 6   - 7   - 7   + 18   + 17

17. 9   18. 13   19. 19   20. 15   21. 7   22. 7   23. 17   24. 1
    - 4   + 2   - 1   - 4   - 3   + 7   - 14   - 1

25. 12   26. 15   27. 9   28. 14   29. 2   30. 12   31. 7   32. 16
    - 5   + 11   - 2   + 3   + 11   - 1   - 4   + 13

33. 9   34. 3   35. 12   36. 14   37. 13   38. 8   39. 7   40. 18
    + 17   + 7   - 2   - 8   + 15   - 6   - 6   - 3

41. 8   42. 9   43. 8   44. 8   45. 8   46. 6   47. 2   48. 12
    + 1   - 3   - 2   - 1   + 7   + 11   - 2   + 10

49. 9   50. 10   51. 14   52. 12   53. 19   54. 5   55. 16   56. 13
    + 19   - 10   + 10   + 8   + 20   + 1   + 6   + 13

57. 14   58. 8   59. 5   60. 6
    - 4   + 14   + 3   - 6

Score
/60

I Feel:

| 1. 20 + 8 | 2. 2 − 2 | 3. 3 + 1 | 4. 18 − 11 | 5. 11 + 15 | 6. 7 − 5 | 7. 7 + 2 | 8. 5 + 15 |

| 9. 18 + 15 | 10. 12 + 19 | 11. 7 − 3 | 12. 17 − 11 | 13. 19 − 18 | 14. 12 + 8 | 15. 10 − 1 | 16. 18 + 13 |

| 17. 3 − 2 | 18. 4 + 16 | 19. 16 − 5 | 20. 9 + 11 | 21. 15 + 10 | 22. 8 − 6 | 23. 8 + 17 | 24. 4 − 3 |

| 25. 11 + 6 | 26. 11 + 10 | 27. 14 − 4 | 28. 5 + 3 | 29. 13 − 6 | 30. 4 + 7 | 31. 9 + 1 | 32. 8 − 4 |

| 33. 18 − 12 | 34. 9 − 0 | 35. 19 − 13 | 36. 10 + 6 | 37. 17 + 14 | 38. 17 + 12 | 39. 19 + 19 | 40. 18 − 7 |

| 41. 0 − 0 | 42. 12 − 8 | 43. 14 − 10 | 44. 14 + 16 | 45. 10 − 3 | 46. 7 − 7 | 47. 3 + 16 | 48. 4 + 15 |

| 49. 14 − 12 | 50. 5 − 2 | 51. 17 − 7 | 52. 9 + 16 | 53. 17 + 17 | 54. 9 − 8 | 55. 18 − 4 | 56. 9 + 6 |

| 57. 2 − 1 | 58. 1 − 0 | 59. 19 + 17 | 60. 19 + 2 |

Score /60

I Feel: ☹ 😐 🙂

Mixed Problems

Start   Finish

Date: ....   NAME : ....   Time: ....:.... ....:....

1. 4
  + 15

2. 9
  - 8

3. 3
  - 2

4. 14
  - 12

5. 19
  + 2

6. 2
  - 1

7. 18
  + 13

8. 18
  - 12

9. 19
  + 19

10. 18
  + 15

11. 17
  + 12

12. 19
  + 17

13. 18
  - 11

14. 7
  - 3

15. 5
  + 15

16. 18
  - 4

17. 4
  - 3

18. 14
  - 10

19. 9
  - 0

20. 14
  + 16

21. 0
  - 0

22. 9
  + 16

23. 15
  + 10

24. 8
  - 6

25. 11
  + 15

26. 5
  - 2

27. 13
  - 6

28. 2
  - 2

29. 18
  - 7

30. 10
  - 3

31. 17
  + 14

32. 12
  + 8

33. 7
  - 7

34. 11
  + 10

35. 9
  + 1

36. 7
  - 5

37. 11
  + 6

38. 4
  + 16

39. 7
  + 2

40. 19
  - 18

41. 8
  + 17

42. 8
  - 4

43. 10
  - 1

44. 1
  - 0

45. 9
  + 6

46. 4
  + 7

47. 9
  + 11

48. 17
  - 7

49. 3
  + 16

50. 17
  - 11

51. 10
  + 6

52. 3
  + 1

53. 12
  - 8

54. 16
  - 5

55. 17
  + 17

56. 19
  - 13

57. 5
  + 3

58. 20
  + 8

59. 12
  + 19

60. 14
  - 4

Score
/60

I Feel: :(:| :)

1. 6 − 5	2. 5 − 3	3. 11 − 1	4. 14 + 13	5. 13 − 4	6. 18 + 18	7. 4 + 14	8. 0 + 3
9. 10 − 4	10. 8 + 11	11. 15 − 8	12. 2 − 1	13. 19 + 18	14. 3 − 0	15. 10 − 10	16. 1 + 9
17. 6 + 14	18. 14 + 15	19. 16 − 11	20. 10 − 6	21. 18 + 15	22. 19 + 14	23. 15 − 0	24. 17 − 10
25. 8 + 4	26. 14 − 12	27. 19 − 4	28. 6 + 10	29. 11 + 5	30. 15 + 10	31. 10 + 16	32. 18 − 1
33. 9 − 1	34. 15 − 14	35. 5 + 16	36. 16 + 2	37. 20 − 2	38. 0 + 12	39. 7 − 4	40. 1 + 5
41. 19 + 5	42. 16 − 14	43. 14 + 11	44. 7 − 5	45. 16 − 15	46. 8 − 2	47. 1 + 11	48. 5 − 2
49. 17 + 3	50. 5 − 5	51. 10 − 1	52. 18 + 4	53. 5 − 4	54. 15 − 2	55. 4 + 2	56. 3 + 5
57. 6 + 4	58. 2 + 1	59. 16 + 5	60. 2 − 2				

Score /60

I Feel: ☹ 😐 🙂

1. 4
   - 3

2. 6
   + 3

3. 16
   - 12

4. 9
   + 19

5. 15
   - 9

6. 17
   + 17

7. 8
   + 15

8. 13
   + 14

9. 5
   + 4

10. 15
    - 6

11. 17
    - 9

12. 12
    - 11

13. 14
    + 13

14. 8
    - 5

15. 4
    + 16

16. 5
    - 4

17. 5
    - 3

18. 14
    + 8

19. 15
    - 10

20. 7
    + 4

21. 16
    - 3

22. 14
    - 9

23. 9
    - 3

24. 6
    + 10

25. 2
    + 12

26. 11
    - 3

27. 3
    - 3

28. 13
    + 6

29. 5
    + 14

30. 4
    - 4

31. 5
    + 15

32. 17
    - 5

33. 12
    + 12

34. 8
    + 20

35. 10
    - 8

36. 18
    + 12

37. 6
    - 4

38. 2
    + 9

39. 11
    + 16

40. 2
    - 2

41. 13
    + 8

42. 16
    - 11

43. 14
    - 6

44. 10
    - 7

45. 4
    + 12

46. 9
    + 2

47. 18
    + 10

48. 14
    - 11

49. 17
    + 15

50. 10
    - 5

51. 19
    - 17

52. 13
    + 16

53. 4
    - 2

54. 3
    + 17

55. 8
    + 13

56. 19
    - 5

57. 7
    - 5

58. 7
    - 6

59. 15
    + 14

60. 11
    + 19

Score
/60

I Feel: 😟 😐 🙂

Mixed Problems

Date: ....   NAME : ....   Start   Finish

Time: ....:.... ....:....

1. 3
   - 1

2. 4
   - 2

3. 18
   - 10

4. 1
   - 1

5. 10
   + 11

6. 9
   - 4

7. 4
   + 16

8. 3
   + 1

9. 17
   - 4

10. 12
    - 6

11. 7
    + 7

12. 17
    - 13

13. 9
    + 1

14. 19
    + 19

15. 16
    - 10

16. 3
    - 3

17. 11
    + 7

18. 14
    - 14

19. 0
    + 15

20. 12
    - 2

21. 13
    - 10

22. 14
    + 7

23. 13
    + 10

24. 1
    + 5

25. 0
    - 0

26. 17
    + 15

27. 10
    + 7

28. 1
    - 0

29. 4
    + 4

30. 17
    + 9

31. 2
    + 0

32. 6
    + 11

33. 11
    - 6

34. 7
    + 10

35. 18
    - 14

36. 18
    + 9

37. 20
    - 8

38. 16
    + 3

39. 13
    + 6

40. 2
    - 2

41. 8
    + 4

42. 12
    + 11

43. 13
    + 4

44. 12
    + 14

45. 7
    - 0

46. 10
    - 0

47. 15
    - 2

48. 16
    + 6

49. 0
    + 20

50. 11
    + 19

51. 17
    - 9

52. 14
    - 7

53. 5
    - 1

54. 19
    + 7

55. 14
    - 0

56. 16
    - 1

57. 11
    - 8

58. 12
    - 1

59. 11
    + 20

60. 19
    - 6

Score
/60

I Feel: ☹ 😐 ☺

Mixed Problems

Date: ....   NAME : ....   Time: ....:.... ....:....

Start   Finish

1.	2.	3.	4.	5.	6.	7.	8.
15	4	13	6	10	15	6	18
+ 1	- 3	+ 14	+ 13	- 1	- 6	+ 16	+ 3

9.	10.	11.	12.	13.	14.	15.	16.
8	17	9	1	15	4	11	4
- 2	- 14	+ 12	+ 19	- 9	+ 6	- 7	+ 19

17.	18.	19.	20.	21.	22.	23.	24.
18	12	17	15	18	9	5	5
+ 9	- 4	- 6	- 12	- 17	+ 8	+ 9	+ 1

25.	26.	27.	28.	29.	30.	31.	32.
15	9	12	11	13	3	0	13
+ 9	- 3	+ 4	- 6	- 3	- 1	- 0	+ 7

33.	34.	35.	36.	37.	38.	39.	40.
8	17	13	11	5	8	19	4
- 1	+ 14	- 2	+ 19	+ 6	- 4	+ 17	- 4

41.	42.	43.	44.	45.	46.	47.	48.
10	16	16	12	11	16	6	11
- 3	- 16	+ 17	+ 19	- 2	- 7	+ 19	+ 13

49.	50.	51.	52.	53.	54.	55.	56.
15	16	1	17	14	7	8	20
+ 6	+ 6	+ 6	+ 4	- 7	- 1	+ 5	- 5

57.	58.	59.	60.
18	5	15	2
+ 17	- 1	- 3	- 1

Score /60

I Feel: ☹ 😐 🙂

| 1. | 19
+ 5 | 2. | 10
- 4 | 3. | 1
+ 14 | 4. | 8
+ 20 | 5. | 2
- 1 | 6. | 16
+ 2 | 7. | 14
- 1 | 8. | 20
- 18 |

9. 13   10. 6   11. 17   12. 4   13. 10   14. 9   15. 3   16. 15
  - 13   + 14   - 15   - 3   + 4   + 4   + 19   + 7

17. 17   18. 6   19. 5   20. 11   21. 15   22. 5   23. 11   24. 3
  + 4   - 5   - 5   - 10   - 7   + 18   - 4   - 2

25. 9   26. 6   27. 15   28. 7   29. 10   30. 11   31. 10   32. 8
  - 5   - 4   - 8   + 10   + 13   - 7   + 8   - 1

33. 14   34. 13   35. 19   36. 1   37. 16   38. 1   39. 4   40. 12
  - 5   - 4   - 14   + 1   - 11   + 10   + 11   - 7

41. 11   42. 2   43. 4   44. 19   45. 5   46. 10   47. 6   48. 8
  + 4   + 18   + 1   + 8   + 7   - 9   + 12   - 3

49. 3   50. 14   51. 11   52. 5   53. 8   54. 13   55. 2   56. 15
  + 11   + 13   + 15   - 3   + 19   - 9   + 2   - 6

57. 18   58. 18   59. 10   60. 4
  - 13   + 11   - 3   + 14

Score
/60

I Feel: ☹ 😐 🙂

Mixed Problems

Date: ....   NAME : ....

Start   Finish

Time: ....:.... ....:....

1.	15	2.	6	3.	7	4.	15	5.	6	6.	9	7.	10	8.	11
	- 8		- 2		- 2		+ 14		+ 18		+ 7		+ 4		- 10

9.	20	10.	6	11.	15	12.	18	13.	6	14.	6	15.	13	16.	8
	+ 2		+ 16		- 11		- 18		+ 12		+ 2		+ 15		+ 6

17.	15	18.	14	19.	13	20.	17	21.	17	22.	2	23.	12	24.	9
	+ 9		+ 2		- 11		+ 9		- 14		- 1		+ 14		- 5

25.	8	26.	20	27.	10	28.	3	29.	4	30.	10	31.	11	32.	4
	+ 16		- 2		- 5		- 3		- 3		+ 20		- 7		- 2

33.	5	34.	16	35.	4	36.	12	37.	12	38.	20	39.	5	40.	9
	+ 3		- 7		- 1		- 9		+ 15		+ 14		- 5		+ 8

41.	12	42.	13	43.	11	44.	13	45.	17	46.	1	47.	20	48.	1
	- 10		+ 17		+ 17		- 8		- 3		+ 10		- 9		+ 1

49.	13	50.	7	51.	14	52.	1	53.	18	54.	19	55.	5	56.	10
	- 2		- 5		+ 7		- 1		+ 9		+ 17		+ 7		+ 1

57.	5	58.	19	59.	12	60.	9
	- 3		- 15		+ 6		- 2

Score
/60

I Feel:

1.   9
  - 7
	2. 5
	  - 3
		3. 8
		  - 2
			4. 1
			  - 1
				5. 5
				  - 1
					6. 16
					 + 10
						7. 15
						 + 8
							8. 12
							 + 4

9. 17 + 18	10. 19 + 9	11. 8 + 8	12. 5 - 2	13. 3 - 2	14. 14 + 12	15. 17 + 8	16. 18 + 13
17. 15 - 13	18. 16 - 14	19. 18 - 2	20. 12 + 19	21. 18 - 6	22. 12 - 4	23. 4 + 19	24. 15 - 1
25. 8 - 8	26. 2 + 10	27. 4 + 1	28. 15 + 1	29. 11 - 1	30. 14 + 9	31. 1 + 5	32. 10 - 5
33. 8 + 3	34. 4 - 3	35. 16 - 8	36. 7 + 15	37. 18 - 4	38. 4 + 14	39. 11 - 3	40. 11 + 18
41. 7 - 2	42. 13 - 2	43. 5 - 5	44. 5 + 7	45. 15 - 6	46. 3 + 4	47. 12 + 3	48. 10 - 8
49. 3 + 9	50. 10 + 8	51. 14 + 1	52. 18 + 4	53. 1 + 18	54. 8 + 5	55. 15 - 4	56. 14 - 3
57. 19 - 6	58. 3 - 1	59. 4 + 10	60. 2 - 1				

Score
/60

I Feel: :(:| :)

1.	2.	3.	4.	5.	6.	7.	8.
8 + 15	7 - 2	3 + 1	8 + 4	8 + 9	12 - 5	4 + 0	7 - 1

9.	10.	11.	12.	13.	14.	15.	16.
2 + 11	5 - 3	1 + 14	10 - 5	8 + 12	2 + 3	15 - 7	5 + 3

17.	18.	19.	20.	21.	22.	23.	24.
4 - 1	16 + 16	7 - 3	2 - 2	14 - 10	1 + 12	5 + 7	10 + 15

25.	26.	27.	28.	29.	30.	31.	32.
2 - 1	16 - 8	12 - 2	9 - 8	4 - 3	18 - 8	3 - 2	9 - 3

33.	34.	35.	36.	37.	38.	39.	40.
5 - 2	6 - 4	2 + 19	13 - 3	8 - 6	9 + 15	5 + 0	6 - 5

41.	42.	43.	44.	45.	46.	47.	48.
6 - 3	10 - 2	10 + 17	3 + 14	9 + 16	20 + 15	11 + 11	12 + 16

49.	50.	51.	52.	53.	54.	55.	56.
14 + 16	3 + 0	8 - 8	12 - 1	9 - 2	3 + 4	12 + 12	15 - 4

57.	58.	59.	60.
14 + 11	16 + 1	13 - 4	14 + 13

Score /60

I Feel: 😟 🙂 😊

Day 100

Mixed Problems

Date: ....   NAME : ....   Start   Finish

Time: ....:.....   ....:.....

1. 16 + 2	2. 12 - 7	3. 19 - 9	4. 7 - 1	5. 15 - 14	6. 10 + 12	7. 16 + 17	8. 18 + 9
9. 5 - 1	10. 18 + 18	11. 17 - 9	12. 8 - 8	13. 3 - 3	14. 14 - 14	15. 13 + 3	16. 9 + 7
17. 7 + 17	18. 8 - 4	19. 3 + 16	20. 16 - 9	21. 8 - 2	22. 20 - 9	23. 4 - 3	24. 3 - 1
25. 19 - 2	26. 17 + 13	27. 20 - 17	28. 6 + 16	29. 16 + 11	30. 19 + 3	31. 5 + 8	32. 10 + 8
33. 8 - 6	34. 11 + 12	35. 11 - 2	36. 15 - 7	37. 10 - 9	38. 11 - 10	39. 8 + 11	40. 3 + 4
41. 11 - 6	42. 20 + 13	43. 9 + 18	44. 11 + 6	45. 3 + 17	46. 7 - 3	47. 17 - 2	48. 3 + 10
49. 7 + 14	50. 0 + 7	51. 18 - 17	52. 8 + 3	53. 16 - 2	54. 13 - 2	55. 9 + 13	56. 8 + 20
57. 12 - 1	58. 6 + 19	59. 13 + 13	60. 10 - 4				

Score /60

I Feel: ☹ 😐 🙂

1.	2.	3.	4.	5.	6.	7.	8.
18 + 63	60 + 35	54 + 69	20 + 63	63 + 52	22 + 39	67 + 58	24 + 90

9.	10.	11.	12.	13.	14.	15.	16.
98 + 11	80 + 55	5 + 81	35 + 22	28 + 59	71 + 8	10 + 27	39 + 15

17.	18.	19.	20.	21.	22.	23.	24.
95 + 51	42 + 96	33 + 82	11 + 81	12 + 9	80 + 84	46 + 43	69 + 77

25.	26.	27.	28.	29.	30.	31.	32.
62 + 82	62 + 23	6 + 73	75 + 35	29 + 37	7 + 94	97 + 13	53 + 36

33.	34.	35.	36.	37.	38.	39.	40.
13 + 7	43 + 17	16 + 27	57 + 16	1 + 98	1 + 50	76 + 84	89 + 93

41.	42.	43.	44.	45.	46.	47.	48.
44 + 28	64 + 16	10 + 22	75 + 6	67 + 94	75 + 16	83 + 98	55 + 44

49.	50.	51.	52.	53.	54.	55.	56.
71 + 39	0 + 52	60 + 36	25 + 46	22 + 7	31 + 78	34 + 78	61 + 22

57.	58.	59.	60.
4 + 3	67 + 25	71 + 96	96 + 21

Score /60

I Feel:

Day 102

BONUS PAGE

Date: ....

Subtracting Digits 0 - 99

NAME : ....

Start   Finish

Time: ....:.... ....:....

1.	2.	3.	4.	5.	6.	7.	8.
94	62	40	43	77	79	76	29
- 20	- 42	- 10	- 11	- 22	- 45	- 4	- 15

9.	10.	11.	12.	13.	14.	15.	16.
85	65	61	54	52	3	16	77
- 54	- 29	- 59	- 7	- 11	- 0	- 15	- 1

17.	18.	19.	20.	21.	22.	23.	24.
66	71	15	57	68	53	75	28
- 7	- 24	- 3	- 32	- 45	- 32	- 33	- 6

25.	26.	27.	28.	29.	30.	31.	32.
19	88	67	53	14	65	26	93
- 10	- 71	- 6	- 48	- 1	- 63	- 22	- 64

33.	34.	35.	36.	37.	38.	39.	40.
20	68	51	62	10	36	3	96
- 1	- 64	- 15	- 13	- 8	- 9	- 1	- 73

41.	42.	43.	44.	45.	46.	47.	48.
6	24	92	1	86	19	2	60
- 4	- 2	- 58	- 1	- 50	- 13	- 0	- 19

49.	50.	51.	52.	53.	54.	55.	56.
29	40	64	64	4	99	86	47
- 10	- 19	- 62	- 6	- 3	- 10	- 42	- 44

57.	58.	59.	60.
27	8	47	45
- 14	- 8	- 21	- 20

Score /60

I Feel: :(:| :)

1. 69
    - 51

2. 48
    + 60

3. 11
    - 8

4. 39
    + 37

5. 37
    - 33

6. 95
    + 78

7. 55
    + 5

8. 14
    + 69

9. 29
    + 59

10. 33
    + 83

11. 26
    - 15

12. 99
    + 10

13. 11
    + 34

14. 25
    - 25

15. 68
    - 64

16. 53
    - 35

17. 33
    + 66

18. 39
    + 85

19. 55
    + 73

20. 37
    - 23

21. 61
    + 95

22. 70
    - 57

23. 9
    + 81

24. 71
    - 54

25. 93
    + 38

26. 2
    - 1

27. 33
    - 32

28. 75
    - 21

29. 94
    - 47

30. 71
    - 68

31. 33
    + 40

32. 33
    + 74

33. 58
    + 99

34. 40
    + 17

35. 94
    - 71

36. 80
    + 67

37. 41
    - 33

38. 29
    + 77

39. 14
    + 73

40. 2
    + 85

41. 93
    + 83

42. 97
    - 54

43. 48
    + 95

44. 67
    - 26

45. 36
    - 5

46. 59
    - 21

47. 58
    - 28

48. 26
    - 18

49. 51
    - 41

50. 46
    - 20

51. 91
    - 21

52. 62
    + 5

53. 45
    + 82

54. 93
    - 92

55. 97
    - 43

56. 40
    - 5

57. 4
    + 62

58. 99
    - 9

59. 5
    + 65

60. 75
    + 99

Score /60

I Feel:

1. 74 − 39	2. 60 − 56	3. 80 + 43	4. 89 + 98	5. 59 + 49	6. 93 + 41	7. 7 − 0	8. 20 + 6
9. 50 + 39	10. 11 + 91	11. 31 − 23	12. 32 − 11	13. 90 + 95	14. 17 + 40	15. 16 + 37	16. 41 + 38
17. 40 − 28	18. 37 + 66	19. 76 + 81	20. 88 + 14	21. 34 − 10	22. 6 − 5	23. 45 − 5	24. 91 + 17
25. 14 + 47	26. 55 − 28	27. 49 − 5	28. 83 − 15	29. 38 + 33	30. 27 − 25	31. 36 − 36	32. 60 − 8
33. 18 − 8	34. 9 − 1	35. 75 − 26	36. 22 − 5	37. 0 − 0	38. 14 + 2	39. 2 + 39	40. 75 + 59
41. 65 + 39	42. 4 + 12	43. 13 + 10	44. 83 − 19	45. 67 + 60	46. 2 + 37	47. 76 + 33	48. 38 + 6
49. 92 − 6	50. 15 − 3	51. 94 + 45	52. 6 − 0	53. 64 − 50	54. 33 − 15	55. 96 + 49	56. 98 + 9
57. 23 − 23	58. 25 − 11	59. 18 − 15	60. 45 − 26				

Score /60

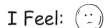

I Feel:

Answer Key

Day 1
1. 7　2. 9　3. 3　4. 3　5. 5　6. 2　7. 2　8. 3　9. 3　10. 8　11. 1　12. 8　13. 1　14. 7　15. 6　16. 4　17. 4　18. 4　19. 6　20. 6
21. 8　22. 5　23. 10　24. 3　25. 4　26. 4　27. 7　28. 3　29. 5　30. 5　31. 6　32. 4　33. 4　34. 3　35. 8　36. 5　37. 5　38. 0　39. 6　40. 9
41. 9　42. 6　43. 5　44. 6　45. 7　46. 3　47. 6　48. 8　49. 5　50. 2　51. 6　52. 7　53. 5　54. 6　55. 6　56. 6　57. 5　58. 1　59. 5　60. 4

Day 2
1. 4　2. 1　3. 3　4. 9　5. 7　6. 2　7. 5　8. 7　9. 5　10. 5　11. 4　12. 5　13. 5　14. 2　15. 8　16. 7　17. 5　18. 9　19. 3　20. 4
21. 7　22. 6　23. 3　24. 2　25. 9　26. 1　27. 3　28. 3　29. 6　30. 6　31. 4　32. 7　33. 4　34. 4　35. 7　36. 8　37. 5　38. 5　39. 5　40. 7
41. 6　42. 4　43. 0　44. 6　45. 6　46. 6　47. 3　48. 5　49. 6　50. 4　51. 5　52. 10　53. 3　54. 4　55. 3　56. 6　57. 7　58. 8　59. 10　60. 8

Day 3
1. 3　2. 10　3. 6　4. 7　5. 7　6. 7　7. 5　8. 7　9. 6　10. 6　11. 3　12. 4　13. 5　14. 9　15. 4　16. 8　17. 4　18. 4　19. 3　20. 5
21. 5　22. 4　23. 5　24. 2　25. 6　26. 6　27. 2　28. 7　29. 6　30. 4　31. 5　32. 3　33. 7　34. 5　35. 6　36. 6　37. 10　38. 5　39. 1　40. 6
41. 3　42. 8　43. 8　44. 0　45. 3　46. 8　47. 5　48. 3　49. 7　50. 7　51. 2　52. 4　53. 1　54. 3　55. 5　56. 9　57. 4　58. 4　59. 5　60. 9

Day 4
1. 6　2. 8　3. 9　4. 6　5. 3　6. 6　7. 3　8. 6　9. 8　10. 4　11. 7　12. 4　13. 5　14. 7　15. 5　16. 3　17. 4　18. 7　19. 5　20. 6
21. 3　22. 4　23. 10　24. 8　25. 2　26. 5　27. 5　28. 6　29. 4　30. 6　31. 7　32. 7　33. 4　34. 3　35. 0　36. 4　37. 5　38. 7　39. 5　40. 3
41. 10　42. 1　43. 9　44. 4　45. 5　46. 7　47. 2　48. 8　49. 5　50. 7　51. 5　52. 4　53. 3　54. 9　55. 3　56. 1　57. 6　58. 6　59. 5　60. 4

Day 5
1. 7　2. 10　3. 5　4. 2　5. 12　6. 9　7. 11　8. 8　9. 8　10. 7　11. 6　12. 1　13. 5　14. 6　15. 6　16. 12　17. 11　18. 3　19. 10　20. 3
21. 7　22. 13　23. 7　24. 5　25. 2　26. 2　27. 14　28. 8　29. 11　30. 4　31. 7　32. 3　33. 9　34. 6　35. 10　36. 7　37. 10　38. 4　39. 9　40. 6
41. 4　42. 4　43. 7　44. 2　45. 4　46. 9　47. 0　48. 6　49. 1　50. 9　51. 5　52. 8　53. 13　54. 8　55. 5　56. 9　57. 11　58. 3　59. 9　60. 3

Day 6
1. 11　2. 2　3. 4　4. 7　5. 9　6. 9　7. 10　8. 3　9. 10　10. 5　11. 1　12. 5　13. 8　14. 11　15. 7　16. 8　17. 7　18. 8　19. 7　20. 6
21. 3　22. 6　23. 5　24. 7　25. 0　26. 7　27. 5　28. 3　29. 9　30. 9　31. 4　32. 11　33. 8　34. 7　35. 4　36. 14　37. 6　38. 2　39. 12　40. 10
41. 7　42. 3　43. 8　44. 13　45. 11　46. 8　47. 10　48. 6　49. 2　50. 9　51. 4　52. 1　53. 4　54. 5　55. 13　56. 9　57. 6　58. 12　59. 8　60. 6

Day 7
1. 4　2. 7　3. 7　4. 2　5. 9　6. 6　7. 5　8. 8　9. 5　10. 1　11. 3　12. 12　13. 8　14. 5　15. 2　16. 8　17. 4　18. 7　19. 13　20. 5
21. 9　22. 3　23. 10　24. 5　25. 7　26. 8　27. 11　28. 1　29. 4　30. 10　31. 8　32. 7　33. 4　34. 4　35. 3　36. 7　37. 0　38. 8　39. 9　40. 2
41. 12　42. 13　43. 11　44. 8　45. 6　46. 11　47. 7　48. 7　49. 6　50. 5　51. 6　52. 3　53. 9　54. 6　55. 10　56. 9　57. 10　58. 6　59. 11　60. 14

Day 8
1. 7　2. 11　3. 5　4. 10　5. 14　6. 6　7. 3　8. 2　9. 7　10. 4　11. 12　12. 7　13. 8　14. 10　15. 7　16. 4　17. 10　18. 6　19. 9　20. 8
21. 12　22. 2　23. 9　24. 13　25. 6　26. 9　27. 8　28. 3　29. 10　30. 7　31. 3　32. 10　33. 9　34. 11　35. 4　36. 8　37. 11　38. 10　39. 10　40. 8
41. 6　42. 7　43. 4　44. 5　45. 6　46. 5　47. 7　48. 5　49. 5　50. 9　51. 13　52. 11　53. 4　54. 6　55. 9　56. 1　57. 8　58. 12　59. 6　60. 5

Day 9
1. 11　2. 12　3. 7　4. 6　5. 10　6. 1　7. 8　8. 9　9. 14　10. 7　11. 5　12. 11　13. 3　14. 11　15. 5　16. 4　17. 6　18. 8　19. 10　20. 4
21. 7　22. 8　23. 8　24. 12　25. 6　26. 8　27. 6　28. 9　29. 5　30. 12　31. 10　32. 8　33. 9　34. 2　35. 7　36. 4　37. 10　38. 3　39. 4　40. 9
41. 7　42. 8　43. 7　44. 5　45. 9　46. 8　47. 6　48. 6　49. 9　50. 6　51. 12　52. 10　53. 13　54. 7　55. 2　56. 11　57. 11　58. 11　59. 5　60. 13

Day 10
1. 7　2. 11　3. 6　4. 5　5. 2　6. 8　7. 13　8. 2　9. 8　10. 5　11. 7　12. 1　13. 3　14. 9　15. 6　16. 9　17. 11　18. 7　19. 10　20. 2
21. 8　22. 7　23. 4　24. 12　25. 6　26. 13　27. 3　28. 7　29. 8　30. 7　31. 6　32. 10　33. 5　34. 3　35. 11　36. 4　37. 11　38. 12　39. 7　40. 14
41. 7　42. 8　43. 3　44. 10　45. 10　46. 9　47. 6　48. 5　49. 9　50. 9　51. 8　52. 4　53. 4　54. 9　55. 4　56. 6　57. 6　58. 7　59. 8　60. 10

Day 11
1. 6　2. 8　3. 13　4. 5　5. 4　6. 4　7. 7　8. 13　9. 4　10. 11　11. 13　12. 12　13. 8　14. 3　15. 16　16. 12　17. 11　18. 5　19. 6　20. 13
21. 13　22. 5　23. 11　24. 14　25. 12　26. 10　27. 4　28. 6　29. 14　30. 9　31. 14　32. 5　33. 9　34. 14　35. 15　36. 7　37. 8　38. 10　39. 20　40. 12
41. 13　42. 6　43. 14　44. 10　45. 7　46. 8　47. 17　48. 13　49. 2　50. 7　51. 9　52. 6　53. 6　54. 13　55. 7　56. 5　57. 19　58. 10　59. 16　60. 8

Day 12
1. 15　2. 14　3. 13　4. 9　5. 20　6. 7　7. 12　8. 3　9. 10　10. 13　11. 7　12. 7　13. 12　14. 17　15. 10　16. 8　17. 11　18. 15　19. 7　20. 13
21. 14　22. 16　23. 4　24. 12　25. 11　26. 11　27. 6　28. 6　29. 6　30. 4　31. 5　32. 4　33. 9　34. 13　35. 13　36. 7　37. 2　38. 10　39. 13　40. 19
41. 13　42. 10　43. 9　44. 7　45. 8　46. 5　47. 14　48. 16　49. 8　50. 8　51. 8　52. 12　53. 14　54. 5　55. 6　56. 14　57. 5　58. 10　59. 8　60. 13

Day 13
1. 7　2. 4　3. 10　4. 12　5. 5　6. 15　7. 14　8. 8　9. 9　10. 10　11. 9　12. 17　13. 15　14. 17　15. 12　16. 7　17. 11　18. 16　19. 5　20. 14
21. 15　22. 11　23. 9　24. 11　25. 6　26. 8　27. 8　28. 7　29. 13　30. 7　31. 12　32. 15　33. 3　34. 10　35. 10　36. 16　37. 7　38. 12　39. 13　40. 12
41. 14　42. 7　43. 13　44. 4　45. 9　46. 3　47. 3　48. 4　49. 10　50. 10　51. 9　52. 10　53. 8　54. 13　55. 11　56. 5　57. 2　58. 18　59. 15　60. 8

Day 14

1.12 2.8 3.8 4.15 5.14 6.17 7.7 8.5 9.13 10.8 11.3 12.7 13.3 14.4 15.15 16.8 17.7 18.9 19.10 20.10
21.12 22.7 23.16 24.9 25.9 26.16 27.17 28.9 29.6 30.18 31.2 32.11 33.8 34.5 35.12 36.12 37.13 38.12 39.7 40.13
41.10 42.3 43.14 44.15 45.4 46.15 47.14 48.10 48.15 50.15 51.10 52.11 53.7 54.11 55.4 56.13 57.8 58.13 59.10 60.11

Day 15

1.15 2.13 3.3 4.13 5.7 6.11 7.17 8.7 9.8 10.16 11.9 12.4 13.13 14.15 15.10 16.13 17.12 18.7 19.17 20.15
21.8 22.14 23.15 24.10 25.14 26.4 27.10 28.15 29.7 30.12 31.9 32.13 33.3 34.15 35.13 36.18 37.12 38.10 39.7 40.6
41.5 42.11 43.8 44.12 45.9 46.8 47.14 48.12 48.4 50.9 51.8 52.8 53.10 54.10 55.16 56.9 57.14 58.11 59.11 60.3

Day 16

1.10 2.9 3.7 4.12 5.10 6.9 7.8 8.6 9.10 10.8 11.7 12.3 13.18 14.11 15.15 16.14 17.10 18.14 19.11 20.13
21.7 22.5 23.15 24.13 25.7 26.15 27.13 28.8 29.10 30.13 31.7 32.11 33.3 34.15 35.12 36.8 37.13 38.4 39.16 40.8
41.15 42.10 43.15 44.9 45.13 46.17 47.8 48.8 48.3 50.11 51.12 52.14 53.16 54.9 55.12 56.12 57.10 58.14 59.9 60.17

Day 17

1.13 2.10 3.9 4.15 5.18 6.14 7.10 8.8 9.11 10.10 11.8 12.12 13.12 14.10 15.17 16.3 17.7 18.10 19.10 20.8
21.4 22.14 23.16 24.5 25.14 26.10 27.11 28.8 29.15 30.7 31.7 32.12 33.12 34.13 35.9 36.9 37.3 38.15 39.8 40.8
41.6 42.13 43.15 44.13 45.15 46.7 47.17 48.13 48.9 50.7 51.12 52.11 53.14 54.9 55.13 56.15 57.16 58.8 59.3 60.11

Day 18

1.14 2.14 3.6 4.8 5.10 6.15 7.11 8.3 9.14 10.9 11.10 12.18 13.14 14.7 15.15 16.3 17.15 18.13 19.12 20.15
21.7 22.13 23.17 24.12 25.12 26.8 27.10 28.8 29.15 30.13 31.3 32.17 33.16 34.11 35.10 36.4 37.9 38.7 39.9 40.13
41.7 42.8 43.12 44.13 45.11 46.10 47.16 48.13 48.9 50.10 51.9 52.8 53.12 54.8 55.10 56.5 57.15 58.8 59.7 60.11

Day 19

1.12 2.14 3.9 4.8 5.9 6.11 7.9 8.16 9.15 10.10 11.8 12.12 13.14 14.10 15.11 16.5 17.8 18.7 19.11 20.7
21.15 22.13 23.10 24.15 25.9 26.8 27.12 28.13 29.8 30.3 31.12 32.14 33.10 34.7 35.15 36.15 37.8 38.4 39.11 40.10
41.18 42.10 43.3 44.10 45.6 46.15 47.3 48.17 48.16 50.13 51.7 52.12 53.5 54.14 55.8 56.17 57.7 58.9 59.13 60.13

Day 20

1.6 2.15 3.9 4.7 5.8 6.3 7.3 8.16 9.10 10.5 11.7 12.18 13.6 14.10 15.6 16.13 17.11 18.4 19.9 20.11
21.8 22.12 23.15 24.17 25.13 26.15 27.10 28.15 29.6 30.10 31.14 32.6 33.16 34.13 35.12 36.10 37.12 38.8 39.8 40.14
41.11 42.7 43.9 44.9 45.14 46.13 47.11 48.1 48.7 50.9 51.2 52.3 53.13 54.13 55.9 56.16 57.17 58.10 59.10 60.3

Day 21

1.25 2.28 3.31 4.34 5.34 6.29 7.33 8.32 9.31 10.30 11.32 12.30 13.34 14.27 15.32 16.37 17.29 18.23 19.29 20.31
21.36 22.31 23.30 24.31 25.35 26.33 27.35 28.29 29.32 30.24 31.32 32.27 33.34 34.33 35.29 36.33 37.27 38.27 39.33 40.24
41.29 42.36 43.32 44.32 45.39 46.30 47.36 48.23 48.27 50.23 51.36 52.35 53.28 54.26 55.28 56.24 57.37 58.39 59.31 60.34

Day 22

1.34 2.31 3.36 4.33 5.28 6.23 7.33 8.39 9.30 10.29 11.29 12.36 13.37 14.31 15.32 16.39 17.27 18.33 19.24 20.33
21.29 22.32 23.27 24.33 25.32 26.29 27.31 28.32 29.23 30.32 31.30 32.29 33.28 34.34 35.37 36.23 37.36 38.26 39.34 40.31
41.29 42.32 43.24 44.31 45.27 46.25 47.35 48.34 48.34 50.27 51.35 52.36 53.28 54.27 55.31 56.30 57.35 58.30 59.31 60.24

Day 23

1.27 2.36 3.30 4.29 5.31 6.29 7.37 8.34 9.27 10.33 11.36 12.26 13.33 14.33 15.35 16.32 17.35 18.32 19.33 20.28
21.36 22.30 23.31 24.31 25.39 26.30 27.23 28.34 29.31 30.37 31.31 32.24 33.34 34.24 35.31 36.31 37.27 38.29 39.29 40.29
41.32 42.23 43.30 44.28 45.29 46.25 47.32 48.36 48.29 50.27 51.23 52.24 53.32 54.33 55.32 56.34 57.34 58.28 59.39 60.35

Day 24

1.27 2.36 3.37 4.32 5.31 6.24 7.29 8.31 9.35 10.29 11.37 12.34 13.34 14.32 15.29 16.31 17.33 18.32 19.30 20.29
21.33 22.36 23.36 24.27 25.27 26.34 27.30 28.35 29.33 30.39 31.31 32.34 33.31 34.36 35.30 36.30 37.27 38.27 39.33 40.23
41.23 42.29 43.33 44.28 45.31 46.35 47.31 48.24 48.32 50.29 51.28 52.32 53.24 54.39 55.34 56.32 57.26 58.23 59.25 60.28

Day 25

1.23 2.29 3.29 4.29 5.39 6.33 7.24 8.32 9.26 10.33 11.29 12.33 13.24 14.32 15.27 16.29 17.30 18.31 19.30 20.36
21.34 22.32 23.27 24.31 25.35 26.24 27.36 28.28 29.33 30.23 31.34 32.28 33.25 34.32 35.34 36.29 37.37 38.31 39.36 40.31
41.34 42.31 43.35 44.36 45.35 46.27 47.32 48.39 48.30 50.33 51.37 52.34 53.23 54.28 55.31 56.27 57.27 58.31 59.30 60.32

Day 26

1.31 2.27 3.34 4.29 5.36 6.34 7.31 8.30 9.30 10.28 11.29 12.37 13.37 14.24 15.28 16.39 17.35 18.29 19.31 20.33
21.35 22.36 23.33 24.24 25.34 26.32 27.29 28.32 29.27 30.39 31.34 32.31 33.31 34.29 35.34 36.32 37.33 38.30 39.28 40.33
41.31 42.29 43.24 44.33 45.27 46.30 47.23 48.35 48.36 50.23 51.27 52.25 53.36 54.26 55.32 56.23 57.32 58.32 59.31 60.27

Day 27

1.23 2.34 3.36 4.29 5.27 6.28 7.32 8.32 9.26 10.27 11.39 12.24 13.31 14.23 15.30 16.36 17.31 18.27 19.31 20.35
21.35 22.29 23.36 24.37 25.29 26.32 27.33 28.39 29.29 30.32 31.29 32.32 33.24 34.37 35.28 36.33 37.24 38.32 39.31 40.34
41.30 42.25 43.31 44.33 45.29 46.30 47.34 48.27 48.31 50.36 51.33 52.27 53.34 54.31 55.30 56.35 57.28 58.23 59.33 60.34

Day 28

1. 34 2. 33 3. 21 4. 34 5. 29 6. 29 7. 28 8. 30 9. 36 10. 38 11. 31 12. 28 13. 25 14. 36 15. 39 16. 24 17. 28 18. 27 19. 35 20. 30
21. 27 22. 29 23. 26 24. 31 25. 32 26. 25 27. 27 28. 38 29. 33 30. 31 31. 30 32. 37 33. 32 34. 30 35. 34 36. 29 37. 26 38. 36 39. 23 40. 34
41. 33 42. 25 43. 32 44. 30 45. 29 46. 38 47. 28 48. 33 48. 35 50. 28 51. 37 52. 28 53. 32 54. 30 55. 32 56. 27 57. 26 58. 29 59. 27 60. 21

Day 29

1. 21 2. 35 3. 4 4. 17 5. 8 6. 7 7. 4 8. 24 9. 18 10. 18 11. 9 12. 9 13. 20 14. 11 15. 19 16. 20 17. 19 18. 22 19. 14 20. 22
21. 33 22. 10 23. 23 24. 25 25. 15 26. 30 27. 25 28. 7 29. 31 30. 31 31. 10 32. 18 33. 20 34. 33 35. 23 36. 21 37. 10 38. 32 39. 14 40. 24
41. 18 42. 8 43. 12 44. 24 45. 15 46. 16 47. 17 48. 32 48. 23 50. 27 51. 23 52. 27 53. 21 54. 22 55. 16 56. 11 57. 28 58. 33 59. 26 60. 28

Day 30

1. 21 2. 32 3. 31 4. 23 5. 14 6. 21 7. 7 8. 33 9. 20 10. 8 11. 9 12. 24 13. 18 14. 28 15. 19 16. 12 17. 18 18. 30 19. 19 20. 4
21. 25 22. 10 23. 23 24. 32 25. 16 26. 22 27. 14 28. 26 29. 23 30. 4 31. 35 32. 33 33. 16 34. 27 35. 17 36. 22 37. 10 38. 15 39. 7 40. 20
41. 27 42. 10 43. 23 44. 24 45. 20 46. 21 47. 15 48. 28 48. 8 50. 11 51. 18 52. 18 53. 17 54. 31 55. 11 56. 33 57. 9 58. 22 59. 24 60. 25

Day 31

1. 14 2. 4 3. 21 4. 4 5. 28 6. 23 7. 26 8. 16 9. 11 10. 24 11. 7 12. 20 13. 21 14. 18 15. 23 16. 30 17. 32 18. 19 19. 33 20. 12
21. 9 22. 11 23. 18 24. 10 25. 15 26. 7 27. 9 28. 8 29. 23 30. 27 31. 25 32. 31 33. 31 34. 28 35. 32 36. 18 37. 8 38. 25 39. 27 40. 22
41. 23 42. 21 43. 24 44. 35 45. 33 46. 22 47. 20 48. 20 49. 10 50. 19 51. 16 52. 10 53. 14 54. 17 55. 18 56. 22 57. 33 58. 15 59. 17 60. 24

Day 32

1. 21 2. 32 3. 25 4. 14 5. 8 6. 10 7. 24 8. 28 9. 8 10. 22 11. 25 12. 9 13. 30 14. 21 15. 7 16. 19 17. 33 18. 33 19. 15 20. 18
21. 35 22. 22 23. 24 24. 17 25. 19 26. 20 27. 4 28. 15 29. 7 30. 20 31. 20 32. 18 33. 18 34. 33 35. 14 36. 24 37. 23 38. 27 39. 10 40. 23
41. 28 42. 12 43. 27 44. 4 45. 26 46. 23 47. 23 48. 32 48. 10 50. 17 51. 16 52. 11 53. 31 54. 21 55. 16 56. 11 57. 31 58. 22 59. 9 60. 18

Day 33

1. 26 2. 24 3. 18 4. 26 5. 17 6. 16 7. 29 8. 11 9. 31 10. 23 11. 33 12. 13 13. 29 14. 10 15. 19 16. 31 17. 21 18. 11 19. 32 20. 26
21. 20 22. 18 23. 24 24. 23 25. 5 26. 15 27. 18 28. 15 29. 7 30. 12 31. 13 32. 15 33. 19 34. 27 35. 26 36. 8 37. 37 38. 19 39. 10 40. 4
41. 12 42. 34 43. 21 44. 14 45. 20 46. 23 47. 13 48. 16 48. 12 50. 31 51. 15 52. 11 53. 29 54. 32 55. 5 56. 12 57. 36 58. 21 59. 19 60. 16

Day 34

1. 20 2. 10 3. 13 4. 31 5. 5 6. 32 7. 34 8. 21 9. 20 10. 12 11. 18 12. 15 13. 24 14. 29 15. 17 16. 14 17. 23 18. 13 19. 5 20. 21
21. 23 22. 19 23. 33 24. 19 25. 12 26. 37 27. 23 28. 31 29. 15 30. 19 31. 7 32. 16 33. 11 34. 16 35. 29 36. 4 37. 32 38. 15 39. 21 40. 16
41. 18 42. 10 43. 29 44. 12 45. 18 46. 36 47. 26 48. 11 48. 12 50. 26 51. 26 52. 19 53. 13 54. 8 55. 24 56. 27 57. 26 58. 15 59. 11 60. 31

Day 35

1. 10 2. 23 3. 11 4. 31 5. 19 6. 13 7. 11 8. 10 9. 21 10. 26 11. 15 12. 15 13. 19 14. 12 15. 29 16. 13 17. 32 18. 23 19. 31 20. 18
21. 21 22. 34 23. 5 24. 12 25. 33 26. 19 27. 36 28. 26 29. 17 30. 16 31. 24 32. 26 33. 31 34. 7 35. 29 36. 23 37. 29 38. 5 39. 16 40. 20
41. 26 42. 8 43. 18 44. 11 45. 27 46. 14 47. 13 48. 32 48. 24 50. 12 51. 20 52. 21 53. 19 54. 15 55. 4 56. 37 57. 16 58. 12 59. 18 60. 15

Day 36

1. 18 2. 11 3. 21 4. 8 5. 33 6. 26 7. 13 8. 31 9. 11 10. 12 11. 29 12. 34 13. 7 14. 18 15. 27 16. 5 17. 36 18. 26 19. 12 20. 10
21. 15 22. 11 23. 20 24. 19 25. 21 26. 12 27. 18 28. 15 29. 23 30. 19 31. 20 32. 16 33. 12 34. 15 35. 17 36. 15 37. 10 38. 31 39. 19 40. 18
41. 26 42. 4 43. 16 44. 32 45. 37 46. 29 47. 13 48. 16 48. 24 50. 14 51. 26 52. 29 53. 24 54. 5 55. 19 56. 23 57. 31 58. 23 59. 32 60. 21

Day 37

1. 13 2. 26 3. 18 4. 31 5. 23 6. 10 7. 19 8. 19 9. 17 10. 36 11. 13 12. 5 13. 7 14. 12 15. 26 16. 12 17. 11 18. 29 19. 27 20. 16
21. 29 22. 5 23. 12 24. 10 25. 18 26. 11 27. 23 28. 20 29. 15 30. 20 31. 15 32. 21 33. 15 34. 37 35. 4 36. 31 37. 29 38. 13 39. 33 40. 26
41. 14 42. 11 43. 21 44. 19 45. 32 46. 18 47. 19 48. 23 48. 24 50. 26 51. 21 52. 34 53. 16 54. 12 55. 8 56. 32 57. 16 58. 31 59. 24 60. 15

Day 38

1. 23 2. 23 3. 10 4. 15 5. 16 6. 31 7. 19 8. 24 9. 19 10. 5 11. 16 12. 21 13. 12 14. 20 15. 5 16. 36 17. 12 18. 15 19. 29 20. 17
21. 11 22. 26 23. 19 24. 8 25. 33 26. 31 27. 34 28. 29 29. 16 30. 4 31. 18 32. 37 33. 13 34. 11 35. 32 36. 24 37. 26 38. 14 39. 31 40. 18
41. 21 42. 21 43. 7 44. 10 45. 15 46. 12 47. 18 48. 26 48. 29 50. 32 51. 11 52. 13 53. 26 54. 13 55. 19 56. 20 57. 15 58. 12 59. 27 60. 23

Day 39

1. 7 2. 15 3. 26 4. 16 5. 29 6. 12 7. 5 8. 23 9. 10 10. 15 11. 11 12. 19 13. 23 14. 13 15. 17 16. 18 17. 26 18. 26 19. 31 20. 20
21. 18 22. 29 23. 19 24. 12 25. 12 26. 23 27. 24 28. 31 29. 5 30. 27 31. 15 32. 29 33. 19 34. 18 35. 12 36. 11 37. 13 38. 32 39. 13 40. 11
41. 8 42. 14 43. 34 44. 36 45. 31 46. 26 47. 10 48. 16 48. 16 50. 24 51. 32 52. 4 53. 37 54. 21 55. 15 56. 20 57. 21 58. 33 59. 19 60. 21

Day 40

1. 8 2. 16 3. 18 4. 4 5. 32 6. 18 7. 21 8. 24 9. 10 10. 5 11. 15 12. 32 13. 15 14. 12 15. 17 16. 24 17. 18 18. 11 19. 21 20. 12
21. 29 22. 23 23. 13 24. 19 25. 31 26. 26 27. 7 28. 12 29. 26 30. 23 31. 5 32. 19 33. 11 34. 16 35. 19 36. 13 37. 11 38. 12 39. 20 40. 13
41. 19 42. 36 43. 10 44. 37 45. 26 46. 31 47. 34 48. 20 48. 26 50. 33 51. 16 52. 31 53. 15 54. 23 55. 29 56. 27 57. 14 58. 15 59. 29 60. 21

Day 41

1. 1 2. 1 3. 3 4. 1 5. 2 6. 4 7. 3 8. 4 9. 2 10. 3 11. 2 12. 5 13. 2 14. 1 15. 4 16. 2 17. 1 18. 3 19. 1 20. 2
21. 4 22. 1 23. 2 24. 2 25. 2 26. 4 27. 1 28. 4 29. 3 30. 3 31. 1 32. 2 33. 3 34. 5 35. 3 36. 3 37. 1 38. 2 39. 3 40. 3
41. 4 42. 2 43. 2 44. 4 45. 1 46. 4 47. 1 48. 1 48. 1 50. 1 51. 2 52. 2 53. 3 54. 3 55. 2 56. 3 57. 2 58. 3 59. 1 60. 1

Day 42

#	1	2	3	4	5	6	7	8	9	10	11	12	13	14	15	16	17	18	19	20
1–20	3	2	3	2	5	2	3	1	2	3	1	2	1	1	1	3	4	1	3	2
21–40	1	1	1	1	2	4	1	3	3	1	2	2	2	2	3	3	1	2	1	3
41–60	3	1	1	5	5	1	2	2	1	3	1	2	2	1	1	2	4	2	2	1

Day 43

#	1	2	3	4	5	6	7	8	9	10	11	12	13	14	15	16	17	18	19	20
1–20	1	3	1	2	3	4	0	1	1	3	2	3	3	5	2	2	1	1	2	1
21–40	3	2	1	2	2	2	2	3	2	1	3	1	1	1	1	2	0	1	1	3
41–60	3	4	4	2	5	2	3	2	3	5	3	1	2	2	1	1	2	2	1	1

Day 44

#	1	2	3	4	5	6	7	8	9	10	11	12	13	14	15	16	17	18	19	20
1–20	2	3	2	0	5	3	1	2	3	1	1	2	1	2	2	1	4	4	0	3
21–40	2	3	1	1	1	2	1	1	2	1	4	5	1	3	3	1	2	3	3	2
41–60	3	1	1	3	2	2	5	2	2	2	2	2	3	1	3	1	1	2	1	1

Day 45

#	1	2	3	4	5	6	7	8	9	10	11	12	13	14	15	16	17	18	19	20
1–20	2	2	1	2	5	3	2	4	1	1	3	1	3	1	2	1	6	3	6	1
21–40	2	1	5	2	2	6	4	4	2	1	5	5	2	1	1	3	2	5	4	3
41–60	4	3	2	3	1	2	3	4	4	2	2	4	4	2	4	7	2	2	5	6

Day 46

#	1	2	3	4	5	6	7	8	9	10	11	12	13	14	15	16	17	18	19	20
1–20	1	5	2	1	2	3	6	1	2	4	2	2	1	3	2	4	7	2	1	3
21–40	3	2	4	5	1	6	5	2	3	2	1	2	1	1	6	1	4	1	1	2
41–60	1	5	2	1	2	1	3	0	3	6	3	3	1	1	3	3	2	3	2	2

Day 47

#	1	2	3	4	5	6	7	8	9	10	11	12	13	14	15	16	17	18	19	20
1–20	4	6	1	2	5	1	6	4	2	2	7	4	3	2	4	2	5	4	1	6
21–40	2	2	2	2	2	2	3	1	1	1	5	3	1	3	4	2	2	2	4	2
41–60	3	4	5	1	2	1	6	2	3	3	3	5	4	1	5	4	4	5	3	1

Day 48

#	1	2	3	4	5	6	7	8	9	10	11	12	13	14	15	16	17	18	19	20
1–20	3	3	6	5	2	1	2	2	1	6	2	2	1	2	2	6	1	1	5	2
21–40	3	1	3	1	6	1	2	2	3	5	3	3	3	1	3	1	2	3	1	4
41–60	1	2	2	1	3	2	0	4	7	1	1	4	3	1	3	4	1	3	4	2

Day 49

#	1	2	3	4	5	6	7	8	9	10	11	12	13	14	15	16	17	18	19	20
1–20	5	2	1	2	5	3	2	7	3	3	3	1	4	1	1	6	1	0	2	1
21–40	4	1	2	1	2	2	6	3	6	3	3	3	3	2	6	2	2	1	2	5
41–60	3	4	1	4	1	1	1	4	2	2	1	1	3	1	1	3	3	2	5	2

Day 50

#	1	2	3	4	5	6	7	8	9	10	11	12	13	14	15	16	17	18	19	20
1–20	1	2	3	3	2	1	2	6	6	5	5	4	3	2	1	1	2	1	4	2
21–40	4	3	0	5	3	1	1	3	1	3	3	6	2	2	5	1	2	3	1	2
41–60	7	4	0	3	2	2	2	1	4	2	1	1	1	2	6	3	1	2	1	1

Day 51

#	1	2	3	4	5	6	7	8	9	10	11	12	13	14	15	16	17	18	19	20
1–20	2	4	7	8	4	5	2	3	2	8	3	3	7	8	1	2	4	4	5	3
21–40	5	4	1	6	6	5	3	6	1	2	7	3	3	6	2	5	1	9	4	5
41–60	5	7	1	2	1	3	2	1	1	4	2	2	1	7	4	1	6	1	9	3

Day 52

#	1	2	3	4	5	6	7	8	9	10	11	12	13	14	15	16	17	18	19	20
1–20	3	3	1	1	6	2	5	5	7	3	1	3	6	2	2	2	9	4	1	1
21–40	2	2	1	7	7	8	4	1	1	7	3	3	4	4	3	2	4	8	5	9
41–60	1	5	1	6	2	3	6	1	8	4	6	3	4	4	5	5	5	2	7	2

Day 53

#	1	2	3	4	5	6	7	8	9	10	11	12	13	14	15	16	17	18	19	20
1–20	2	5	2	5	1	1	5	8	2	6	8	1	7	4	7	9	3	2	1	5
21–40	9	4	6	4	2	4	4	3	7	3	3	1	2	2	5	5	7	1	6	8
41–60	2	4	3	1	6	4	6	4	3	1	2	1	1	3		5	3	1	3	7

Day 54

#	1	2	3	4	5	6	7	8	9	10	11	12	13	14	15	16	17	18	19	20
1–20	4	3	1	3	1	4	1	3	6	2	5	1	1	8	3	4	2	2	7	7
21–40	5	6	8	5	6	5	6	2	7	9	2	3	3	3	5	1	2	4	2	1
41–60	3	1	1	2	2	5	6	1	1	4	4	9	1	4	4	2	7	1	6	5

Day 55

#	1	2	3	4	5	6	7	8	9	10	11	12	13	14	15	16	17	18	19	20
1–20	1	3	2	1	1	5	4	1	1	3	4	1	6	9	6	2	4	6	5	4
21–40	5	2	4	1	2	1	2	2	6	3	7	3	2	3	3	2	5	4	8	8
41–60	7	6	7	9	1	8	1	1	2	1	2	9	3	4	1	1	2	2	1	4

Day 56

1. 1 2. 8 3. 2 4. 3 5. 3 6. 1 7. 2 8. 4 9. 5 10. 6 11. 4 12. 1 13. 1 14. 4 15. 5 16. 3 17. 1 18. 1 19. 4 20. 2
21. 5 22. 6 23. 7 24. 1 25. 2 26. 7 27. 2 28. 8 29. 2 30. 4 31. 1 32. 3 33. 2 34. 5 35. 1 36. 6 37. 2 38. 1 39. 5 40. 1
41. 3 42. 4 43. 1 44. 7 45. 9 46. 2 47. 3 48. 7 49. 4 50. 1 51. 6 52. 9 53. 3 54. 6 55. 5 56. 2 57. 2 58. 3 59. 3 60. 8

Day 57

1. 3 2. 5 3. 6 4. 8 5. 1 6. 1 7. 2 8. 3 9. 1 10. 6 11. 3 12. 1 13. 1 14. 9 15. 9 16. 2 17. 2 18. 3 19. 1 20. 2
21. 2 22. 1 23. 2 24. 2 25. 2 26. 6 27. 5 28. 8 29. 4 30. 8 31. 5 32. 1 33. 7 34. 7 35. 2 36. 1 37. 1 38. 5 39. 3 40. 3
41. 4 42. 2 43. 1 44. 4 45. 2 46. 3 47. 5 48. 3 49. 5 50. 4 51. 6 52. 7 53. 6 54. 4 55. 7 56. 4 57. 3 58. 1 59. 1 60. 4

Day 58

1. 2 2. 6 3. 7 4. 1 5. 1 6. 1 7. 8 8. 1 9. 3 10. 1 11. 2 12. 1 13. 5 14. 4 15. 3 16. 1 17. 1 18. 1 19. 1 20. 3
21. 5 22. 4 23. 3 24. 8 25. 6 26. 2 27. 2 28. 1 29. 6 30. 6 31. 1 32. 3 33. 5 34. 3 35. 3 36. 2 37. 4 38. 2 39. 8 40. 2
41. 5 42. 6 43. 9 44. 5 45. 3 46. 9 47. 4 48. 7 49. 4 50. 4 51. 3 52. 2 53. 2 54. 2 55. 4 56. 1 57. 7 58. 5 59. 2 60. 7

Day 59

1. 6 2. 9 3. 2 4. 4 5. 3 6. 6 7. 1 8. 1 9. 1 10. 4 11. 2 12. 3 13. 4 14. 2 15. 1 16. 2 17. 5 18. 5 19. 2 20. 6
21. 3 22. 1 23. 3 24. 3 25. 2 26. 5 27. 1 28. 4 29. 8 30. 2 31. 8 32. 1 33. 7 34. 8 35. 4 36. 5 37. 1 38. 2 39. 3 40. 7
41. 4 42. 7 43. 5 44. 1 45. 1 46. 6 47. 6 48. 3 49. 7 50. 2 51. 1 52. 2 53. 9 54. 3 55. 5 56. 1 57. 4 58. 2 59. 3 60. 1

Day 60

1. 3 2. 6 3. 3 4. 5 5. 6 6. 1 7. 5 8. 3 9. 2 10. 7 11. 2 12. 5 13. 2 14. 1 15. 3 16. 2 17. 6 18. 3 19. 9 20. 4
21. 5 22. 3 23. 1 24. 3 25. 7 26. 4 27. 2 28. 9 29. 2 30. 6 31. 4 32. 8 33. 1 34. 4 35. 4 36. 1 37. 1 38. 2 39. 8 40. 2
41. 7 42. 5 43. 8 44. 1 45. 1 46. 7 47. 6 48. 1 49. 4 50. 1 51. 3 52. 2 53. 3 54. 4 55. 5 56. 5 57. 3 58. 4 59. 2 60. 2

Day 61

1. 2 2. 1 3. 7 4. 1 5. 1 6. 1 7. 0 8. 5 9. 0 10. 0 11. 3 12. 2 13. 8 14. 1 15. 5 16. 1 17. 4 18. 5 19. 2 20. 3
21. 2 22. 8 23. 6 24. 3 25. 0 26. 7 27. 2 28. 1 29. 4 30. 9 31. 0 32. 0 33. 0 34. 1 35. 0 36. 0 37. 4 38. 0 39. 2 40. 5
41. 3 42. 7 43. 3 44. 2 45. 7 46. 6 47. 4 48. 5 49. 3 50. 4 51. 6 52. 8 53. 4 54. 1 55. 3 56. 10 57. 4 58. 6 59. 1 60. 0

Day 62

1. 4 2. 0 3. 3 4. 0 5. 0 6. 2 7. 1 8. 1 9. 0 10. 0 11. 2 12. 3 13. 3 14. 4 15. 2 16. 5 17. 1 18. 5 19. 5 20. 4
21. 2 22. 1 23. 1 24. 4 25. 7 26. 2 27. 5 28. 0 29. 3 30. 10 31. 4 32. 8 33. 2 34. 8 35. 5 36. 1 37. 9 38. 2 39. 4 40. 9
41. 2 42. 6 43. 0 44. 6 45. 3 46. 6 47. 7 48. 7 49. 6 50. 4 51. 3 52. 1 53. 7 54. 5 55. 5 56. 6 57. 3 58. 1 59. 0 60. 0

Day 63

1. 3 2. 2 3. 4 4. 9 5. 3 6. 0 7. 1 8. 5 9. 7 10. 1 11. 3 12. 1 13. 6 14. 3 15. 0 16. 2 17. 2 18. 0 19. 4 20. 5
21. 6 22. 0 23. 9 24. 2 25. 3 26. 5 27. 1 28. 7 29. 3 30. 1 31. 5 32. 4 33. 0 34. 8 35. 7 36. 4 37. 3 38. 0 39. 3 40. 1
41. 2 42. 5 43. 0 44. 6 45. 4 46. 2 47. 1 48. 0 49. 4 50. 2 51. 10 52. 6 53. 1 54. 0 55. 8 56. 5 57. 4 58. 7 59. 6 60. 2

Day 64

1. 6 2. 6 3. 4 4. 2 5. 7 6. 0 7. 10 8. 3 9. 0 10. 1 11. 5 12. 6 13. 2 14. 6 15. 1 16. 2 17. 5 18. 2 19. 1 20. 6
21. 4 22. 0 23. 5 24. 7 25. 4 26. 4 27. 0 28. 0 29. 0 30. 1 31. 0 32. 5 33. 9 34. 4 35. 3 36. 5 37. 4 38. 0 39. 1 40. 0
41. 9 42. 3 43. 5 44. 1 45. 2 46. 3 47. 0 48. 1 49. 2 50. 8 51. 2 52. 2 53. 4 54. 3 55. 8 56. 4 57. 3 58. 7 59. 1 60. 7

Day 65

1. 1 2. 5 3. 7 4. 8 5. 6 6. 3 7. 0 8. 3 9. 7 10. 2 11. 5 12. 4 13. 1 14. 2 15. 2 16. 3 17. 3 18. 2 19. 10 20. 7
21. 4 22. 4 23. 0 24. 1 25. 6 26. 2 27. 4 28. 4 29. 0 30. 9 31. 3 32. 1 33. 1 34. 0 35. 6 36. 1 37. 8 38. 2 39. 6 40. 0
41. 4 42. 7 43. 4 44. 3 45. 5 46. 0 47. 0 48. 0 49. 5 50. 6 51. 1 52. 1 53. 3 54. 5 55. 5 56. 2 57. 9 58. 1 59. 0 60. 6

Day 66

1. 2 2. 5 3. 6 4. 4 5. 8 6. 5 7. 7 8. 2 9. 3 10. 2 11. 3 12. 3 13. 4 14. 6 15. 4 16. 4 17. 2 18. 1 19. 1 20. 0
21. 0 22. 6 23. 8 24. 2 25. 2 26. 0 27. 5 28. 0 29. 4 30. 3 31. 2 32. 10 33. 1 34. 0 35. 7 36. 6 37. 4 38. 3 39. 9 40. 1
41. 1 42. 4 43. 7 44. 2 45. 5 46. 1 47. 0 48. 0 49. 0 50. 6 51. 3 52. 9 53. 2 54. 1 55. 1 56. 0 57. 7 58. 5 59. 3 60. 5

Day 67

1. 4 2. 9 3. 3 4. 3 5. 2 6. 4 7. 7 8. 1 9. 8 10. 4 11. 3 12. 0 13. 10 14. 3 15. 7 16. 5 17. 9 18. 7 19. 4 20. 0
21. 4 22. 4 23. 6 24. 6 25. 3 26. 5 27. 1 28. 5 29. 0 30. 1 31. 2 32. 1 33. 0 34. 0 35. 0 36. 1 37. 4 38. 1 39. 6 40. 0
41. 1 42. 5 43. 6 44. 2 45. 1 46. 2 47. 5 48. 3 49. 8 50. 1 51. 2 52. 3 53. 5 54. 0 55. 2 56. 0 57. 2 58. 2 59. 2 60. 5

Day 68

1. 8 2. 0 3. 3 4. 5 5. 1 6. 4 7. 2 8. 3 9. 0 10. 5 11. 1 12. 3 13. 2 14. 2 15. 0 16. 4 17. 1 18. 1 19. 7 20. 3
21. 2 22. 2 23. 6 24. 0 25. 8 26. 2 27. 5 28. 4 29. 5 30. 6 31. 3 32. 3 33. 0 34. 9 35. 2 36. 5 37. 0 38. 8 39. 5 40. 4
41. 3 42. 6 43. 3 44. 0 45. 1 46. 0 47. 2 48. 6 49. 1 50. 1 51. 1 52. 4 53. 7 54. 6 55. 1 56. 7 57. 4 58. 0 59. 4 60. 2

Day 69

1. 16 2. 2 3. 3 4. 7 5. 6 6. 8 7. 10 8. 9 9. 1 10. 5 11. 5 12. 0 13. 8 14. 5 15. 8 16. 0 17. 7 18. 0 19. 0 20. 4
21. 4 22. 5 23. 8 24. 6 25. 12 26. 1 27. 12 28. 4 29. 1 30. 0 31. 9 32. 9 33. 7 34. 8 35. 11 36. 3 37. 2 38. 5 39. 14 40. 5
41. 7 42. 6 43. 7 44. 4 45. 15 46. 5 47. 1 48. 7 49. 2 50. 14 51. 10 52. 4 53. 2 54. 14 55. 11 56. 3 57. 3 58. 15 59. 0 60. 4

Day 70

1. 11 2. 3 3. 5 4. 8 5. 7 6. 12 7. 9 8. 7 9. 4 10. 4 11. 4 12. 4 13. 8 14. 0 15. 2 16. 0 17. 14 18. 10 19. 3 20. 5
21. 9 22. 5 23. 1 24. 1 25. 0 26. 8 27. 3 28. 15 29. 11 30. 7 31. 6 32. 0 33. 14 34. 2 35. 4 36. 15 37. 4 38. 5 39. 3 40. 0
41. 6 42. 9 43. 14 44. 12 45. 2 46. 5 47. 6 48. 5 48. 2 50. 16 51. 5 52. 1 53. 7 54. 0 55. 8 56. 10 57. 8 58. 1 59. 7 60. 7

Day 71

1. 14 2. 15 3. 8 4. 6 5. 9 6. 0 7. 5 8. 5 9. 7 10. 1 11. 10 12. 1 13. 4 14. 8 15. 5 16. 1 17. 2 18. 12 19. 14 20. 9
21. 14 22. 15 23. 3 24. 6 25. 2 26. 2 27. 3 28. 5 29. 1 30. 7 31. 0 32. 4 33. 5 34. 6 35. 3 36. 4 37. 0 38. 7 39. 4 40. 8
41. 0 42. 16 43. 5 44. 12 45. 2 46. 4 47. 11 48. 4 48. 8 50. 0 51. 3 52. 7 53. 9 54. 5 55. 10 56. 7 57. 7 58. 0 59. 11 60. 8

Day 72

1. 3 2. 3 3. 15 4. 12 5. 9 6. 5 7. 2 8. 6 9. 2 10. 1 11. 4 12. 9 13. 2 14. 0 15. 12 16. 4 17. 3 18. 5 19. 0 20. 1
21. 10 22. 7 23. 9 24. 5 25. 2 26. 13 27. 0 28. 2 29. 14 30. 7 31. 3 32. 2 33. 1 34. 1 35. 11 36. 0 37. 10 38. 0 39. 7 40. 13
41. 5 42. 1 43. 8 44. 3 45. 2 46. 0 47. 0 48. 1 48. 3 50. 7 51. 5 52. 7 53. 5 54. 6 55. 7 56. 9 57. 17 58. 8 59. 11 60. 16

Day 73

1. 9 2. 5 3. 5 4. 5 5. 9 6. 1 7. 2 8. 4 9. 5 10. 17 11. 2 12. 4 13. 7 14. 0 15. 11 16. 1 17. 3 18. 0 19. 1 20. 3
21. 7 22. 0 23. 8 24. 6 25. 2 26. 7 27. 7 28. 10 29. 1 30. 10 31. 14 32. 4 33. 5 34. 12 35. 8 36. 2 37. 15 38. 13 39. 2 40. 7
41. 1 42. 2 43. 2 44. 12 45. 9 46. 5 47. 3 48. 0 48. 0 50. 1 51. 11 52. 16 53. 0 54. 3 55. 6 56. 0 57. 3 58. 9 59. 0 60. 13

Day 74

1. 4 2. 1 3. 0 4. 2 5. 1 6. 10 7. 13 8. 3 9. 2 10. 1 11. 2 12. 3 13. 6 14. 1 15. 2 16. 11 17. 2 18. 15 19. 13 20. 7
21. 0 22. 4 23. 3 24. 8 25. 12 26. 6 27. 7 28. 10 29. 7 30. 3 31. 5 32. 4 33. 1 34. 0 35. 4 36. 2 37. 9 38. 1 39. 0 40. 7
41. 5 42. 11 43. 2 44. 17 45. 0 46. 9 47. 9 48. 12 48. 3 50. 5 51. 14 52. 7 53. 0 54. 8 55. 5 56. 5 57. 5 58. 0 59. 16 60. 9

Day 75

1. 8 2. 1 3. 5 4. 2 5. 11 6. 0 7. 7 8. 2 9. 8 10. 3 11. 11 12. 0 13. 0 14. 5 15. 13 16. 0 17. 10 18. 3 19. 9 20. 1
21. 2 22. 6 23. 12 24. 10 25. 4 26. 0 27. 15 28. 0 29. 14 30. 0 31. 12 32. 4 33. 2 34. 1 35. 9 36. 9 37. 2 38. 5 39. 5 40. 1
41. 6 42. 3 43. 9 44. 5 45. 7 46. 4 47. 3 48. 3 48. 1 50. 5 51. 7 52. 7 53. 7 54. 16 55. 13 56. 2 57. 1 58. 2 59. 3 60. 17

Day 76

1. 12 2. 5 3. 0 4. 9 5. 5 6. 2 7. 7 8. 0 9. 4 10. 6 11. 1 12. 0 13. 7 14. 1 15. 7 16. 7 17. 14 18. 13 19. 11 20. 8
21. 2 22. 0 23. 16 24. 10 25. 2 26. 10 27. 0 28. 3 29. 6 30. 2 31. 3 32. 7 33. 17 34. 1 35. 2 36. 2 37. 1 38. 0 39. 0 40. 1
41. 2 42. 9 43. 4 44. 5 45. 5 46. 3 47. 5 48. 3 48. 3 50. 11 51. 4 52. 9 53. 5 54. 8 55. 3 56. 15 57. 1 58. 9 59. 12 60. 13

Day 77

1. 4 2. 5 3. 6 4. 2 5. 0 6. 17 7. 3 8. 10 9. 1 10. 7 11. 11 12. 10 13. 10 14. 0 15. 8 16. 8 17. 6 18. 0 19. 10 20. 4
21. 5 22. 1 23. 7 24. 13 25. 1 26. 4 27. 2 28. 11 29. 9 30. 13 31. 15 32. 0 33. 5 34. 2 35. 3 36. 5 37. 1 38. 14 39. 4 40. 5
41. 6 42. 14 43. 3 44. 9 45. 12 46. 3 47. 0 48. 5 48. 4 50. 7 51. 9 52. 7 53. 4 54. 0 55. 13 56. 1 57. 7 58. 8 59. 6 60. 5

Day 78

1. 6 2. 11 3. 6 4. 0 5. 0 6. 10 7. 1 8. 5 9. 3 10. 8 11. 14 12. 17 13. 9 14. 5 15. 15 16. 12 17. 13 18. 10 19. 8 20. 4
21. 2 22. 0 23. 7 24. 4 25. 2 26. 11 27. 7 28. 2 29. 4 30. 4 31. 1 32. 1 33. 5 34. 3 35. 0 36. 1 37. 14 38. 7 39. 13 40. 10
41. 7 42. 4 43. 8 44. 6 45. 13 46. 10 47. 5 48. 5 48. 5 50. 7 51. 6 52. 9 53. 0 54. 0 55. 4 56. 5 57. 3 58. 9 59. 0 60. 3

Day 79

1. 7 2. 10 3. 8 4. 4 5. 0 6. 12 7. 0 8. 4 9. 9 10. 2 11. 5 12. 11 13. 5 14. 0 15. 6 16. 10 17. 13 18. 7 19. 4 20. 2
21. 1 22. 7 23. 2 24. 14 25. 1 26. 3 27. 4 28. 5 29. 13 30. 4 31. 13 32. 0 33. 6 34. 9 35. 1 36. 14 37. 11 38. 8 39. 6 40. 1
41. 5 42. 4 43. 3 44. 5 45. 10 46. 3 47. 10 48. 7 48. 15 50. 7 51. 6 52. 9 53. 3 54. 5 55. 8 56. 17 57. 0 58. 5 59. 0 60. 1

Day 80

1. 13 2. 1 3. 1 4. 12 5. 1 6. 8 7. 9 8. 2 9. 4 10. 5 11. 10 12. 3 13. 2 14. 4 15. 1 16. 0 17. 16 18. 8 19. 3 20. 0
21. 0 22. 3 23. 6 24. 4 25. 6 26. 6 27. 3 28. 13 29. 0 30. 14 31. 6 32. 2 33. 6 34. 1 35. 1 36. 4 37. 7 38. 10 39. 5 40. 8
41. 12 42. 4 43. 12 44. 1 45. 7 46. 10 47. 9 48. 8 48. 5 50. 0 51. 1 52. 17 53. 12 54. 4 55. 12 56. 1 57. 0 58. 6 59. 0 60. 11

Day 81

1. 11 2. 14 3. 9 4. 11 5. 23 6. 1 7. 6 8. 22 9. 2 10. 30 11. 0 12. 3 13. 34 14. 23 15. 2 16. 15 17. 4 18. 14 19. 5 20. 11
21. 3 22. 0 23. 29 24. 12 25. 32 26. 13 27. 3 28. 15 29. 32 30. 14 31. 17 32. 9 33. 4 34. 0 35. 26 36. 1 37. 18 38. 27 39. 18 40. 1
41. 3 42. 1 43. 28 44. 4 45. 28 46. 29 47. 27 48. 3 48. 20 50. 4 51. 3 52. 16 53. 2 54. 0 55. 3 56. 24 57. 26 58. 18 59. 11 60. 25

Day 82

1. 0 2. 18 3. 4 4. 2 5. 29 6. 4 7. 11 8. 3 9. 32 10. 0 11. 14 12. 28 13. 11 14. 2 15. 3 16. 6 17. 20 18. 11 19. 4 20. 14
21. 3 22. 0 23. 26 24. 34 25. 27 26. 16 27. 4 28. 15 29. 3 30. 2 31. 0 32. 15 33. 1 34. 27 35. 13 36. 29 37. 32 38. 18 39. 26 40. 11
41. 12 42. 14 43. 17 44. 23 45. 3 46. 3 47. 3 48. 28 48. 5 50. 1 51. 23 52. 1 53. 1 54. 30 55. 25 56. 22 57. 24 58. 9 59. 9 60. 18

Day 83

1. 5 2. 10 3. 9 4. 19 5. 21 6. 20 7. 1 8. 3 9. 30 10. 1 11. 4 12. 14 13. 2 14. 3 15. 1 16. 22 17. 1 18. 7 19. 2 20. 31
21. 12 22. 13 23. 22 24. 3 25. 26 26. 2 27. 1 28. 11 29. 27 30. 28 31. 5 32. 1 33. 6 34. 11 35. 7 36. 21 37. 15 38. 33 39. 5 40. 29
41. 16 42. 10 43. 12 44. 17 45. 2 46. 21 47. 5 48. 32 48. 2 50. 1 51. 9 52. 3 53. 15 54. 8 55. 28 56. 26 57. 12 58. 6 59. 0 60. 35

Day 84
1. 14 2. 9 3. 21 4. 1 5. 3 6. 12 7. 3 8. 28 9. 5 10. 1 11. 1 12. 15 13. 5 14. 12 15. 35 16. 12 17. 15 18. 17 19. 21 20. 1
21. 6 22. 10 23. 1 24. 1 25. 1 26. 22 27. 26 28. 19 29. 2 30. 2 31. 3 32. 5 33. 3 34. 28 35. 2 36. 9 37. 2 38. 27 39. 22 40. 4
41. 11 42. 5 43. 7 44. 31 45. 8 46. 26 47. 0 48. 6 49. 10 50. 13 51. 16 52. 2 53. 21 54. 7 55. 32 56. 33 57. 11 58. 29 59. 30 60. 20

Day 85
1. 22 2. 4 3. 7 4. 37 5. 2 6. 3 7. 2 8. 5 9. 8 10. 1 11. 9 12. 16 13. 28 14. 24 15. 7 16. 15 17. 30 18. 1 19. 7 20. 3
21. 25 22. 14 23. 11 24. 17 25. 17 26. 19 27. 8 28. 13 29. 32 30. 35 31. 6 32. 0 33. 5 34. 28 35. 1 36. 7 37. 4 38. 26 39. 0 40. 25
41. 28 42. 21 43. 5 44. 14 45. 1 46. 1 47. 5 48. 31 49. 15 50. 2 51. 25 52. 9 53. 2 54. 8 55. 14 56. 6 57. 28 58. 19 59. 7 60. 13

Day 86
1. 10 2. 20 3. 5 4. 5 5. 27 6. 11 7. 3 8. 28 9. 3 10. 25 11. 4 12. 1 13. 0 14. 13 15. 7 16. 17 17. 22 18. 19 19. 23 20. 2
21. 5 22. 14 23. 2 24. 0 25. 3 26. 2 27. 18 28. 6 29. 15 30. 0 31. 0 32. 5 33. 17 34. 10 35. 4 36. 1 37. 1 38. 2 39. 29 40. 2
41. 34 42. 28 43. 33 44. 32 45. 35 46. 5 47. 6 48. 12 49. 28 50. 6 51. 37 52. 1 53. 8 54. 24 55. 3 56. 10 57. 9 58. 30 59. 2 60. 23

Day 87
1. 3 2. 11 3. 8 4. 17 5. 33 6. 4 7. 6 8. 1 9. 26 10. 39 11. 7 12. 22 13. 23 14. 13 15. 6 16. 39 17. 26 18. 19 19. 15 20. 28
21. 4 22. 8 23. 24 24. 15 25. 28 26. 5 27. 0 28. 14 29. 0 30. 15 31. 0 32. 7 33. 5 34. 10 35. 9 36. 1 37. 20 38. 27 39. 3 40. 2
41. 1 42. 17 43. 22 44. 24 45. 5 46. 10 47. 0 48. 6 49. 10 50. 7 51. 11 52. 22 53. 6 54. 26 55. 31 56. 29 57. 18 58. 22 59. 0 60. 0

Day 88
1. 7 2. 5 3. 39 4. 1 5. 18 6. 22 7. 28 8. 39 9. 13 10. 20 11. 8 12. 4 13. 29 14. 33 15. 10 16. 22 17. 28 18. 15 19. 3 20. 15
21. 24 22. 0 23. 17 24. 26 25. 26 26. 31 27. 0 28. 24 29. 7 30. 17 31. 27 32. 6 33. 8 34. 6 35. 14 36. 0 37. 15 38. 3 39. 0 40. 1
41. 11 42. 1 43. 5 44. 5 45. 23 46. 22 47. 22 48. 0 49. 2 50. 6 51. 26 52. 19 53. 10 54. 9 55. 6 56. 7 57. 4 58. 11 59. 10 60. 0

Day 89
1. 33 2. 1 3. 24 4. 8 5. 0 6. 31 7. 39 8. 19 9. 5 10. 23 11. 1 12. 5 13. 0 14. 4 15. 27 16. 22 17. 5 18. 15 19. 18 20. 11
21. 4 22. 14 23. 3 24. 0 25. 7 26. 26 27. 7 28. 17 29. 13 30. 11 31. 3 32. 29 33. 26 34. 10 35. 10 36. 6 37. 28 38. 2 39. 1 40. 15
41. 9 42. 6 43. 6 44. 7 45. 15 46. 17 47. 0 48. 22 49. 28 50. 0 51. 24 52. 20 53. 39 54. 6 55. 22 56. 26 57. 10 58. 22 59. 8 60. 0

Day 90
1. 28 2. 0 3. 4 4. 7 5. 26 6. 2 7. 9 8. 20 9. 33 10. 31 11. 4 12. 6 13. 1 14. 20 15. 9 16. 31 17. 1 18. 20 19. 11 20. 20
21. 25 22. 2 23. 25 24. 1 25. 17 26. 21 27. 10 28. 8 29. 7 30. 11 31. 10 32. 4 33. 6 34. 9 35. 6 36. 16 37. 31 38. 29 39. 38 40. 11
41. 0 42. 4 43. 4 44. 30 45. 7 46. 0 47. 19 48. 19 49. 2 50. 3 51. 10 52. 25 53. 34 54. 1 55. 14 56. 15 57. 1 58. 1 59. 36 60. 21

Day 91
1. 19 2. 1 3. 1 4. 2 5. 21 6. 1 7. 31 8. 6 9. 38 10. 33 11. 29 12. 36 13. 7 14. 4 15. 20 16. 14 17. 1 18. 4 19. 9 20. 30
21. 0 22. 25 23. 25 24. 2 25. 26 26. 3 27. 7 28. 0 29. 11 30. 7 31. 31 32. 20 33. 0 34. 21 35. 10 36. 2 37. 17 38. 20 39. 9 40. 1
41. 25 42. 4 43. 9 44. 1 45. 15 46. 11 47. 20 48. 10 49. 19 50. 6 51. 16 52. 4 53. 4 54. 11 55. 34 56. 6 57. 8 58. 28 59. 31 60. 10

Day 92
1. 1 2. 2 3. 10 4. 27 5. 9 6. 36 7. 18 8. 3 9. 6 10. 19 11. 7 12. 1 13. 37 14. 3 15. 0 16. 10 17. 20 18. 29 19. 5 20. 4
21. 33 22. 33 23. 15 24. 7 25. 12 26. 2 27. 15 28. 16 29. 16 30. 25 31. 26 32. 17 33. 8 34. 1 35. 21 36. 18 37. 18 38. 12 39. 3 40. 6
41. 24 42. 2 43. 25 44. 2 45. 1 46. 6 47. 12 48. 3 49. 20 50. 0 51. 9 52. 22 53. 1 54. 13 55. 6 56. 8 57. 10 58. 3 59. 21 60. 0

Day 93
1. 1 2. 9 3. 4 4. 28 5. 6 6. 34 7. 23 8. 27 9. 9 10. 9 11. 8 12. 1 13. 27 14. 3 15. 20 16. 1 17. 2 18. 22 19. 5 20. 11
21. 13 22. 5 23. 6 24. 16 25. 14 26. 8 27. 0 28. 19 29. 19 30. 0 31. 20 32. 12 33. 24 34. 28 35. 2 36. 30 37. 2 38. 11 39. 27 40. 0
41. 21 42. 5 43. 8 44. 3 45. 16 46. 11 47. 28 48. 3 49. 32 50. 5 51. 2 52. 29 53. 2 54. 20 55. 21 56. 14 57. 2 58. 1 59. 29 60. 30

Day 94
1. 2 2. 2 3. 8 4. 0 5. 21 6. 5 7. 20 8. 4 9. 13 10. 6 11. 14 12. 4 13. 10 14. 38 15. 6 16. 0 17. 18 18. 0 19. 15 20. 10
21. 3 22. 21 23. 23 24. 6 25. 0 26. 32 27. 17 28. 1 29. 8 30. 26 31. 2 32. 17 33. 5 34. 17 35. 4 36. 27 37. 12 38. 19 39. 19 40. 0
41. 12 42. 23 43. 17 44. 26 45. 7 46. 10 47. 13 48. 22 49. 20 50. 30 51. 8 52. 7 53. 4 54. 26 55. 14 56. 15 57. 3 58. 11 59. 31 60. 13

Day 95
1. 16 2. 1 3. 27 4. 19 5. 9 6. 9 7. 22 8. 21 9. 6 10. 3 11. 21 12. 20 13. 6 14. 10 15. 4 16. 23 17. 27 18. 8 19. 11 20. 3
21. 1 22. 17 23. 14 24. 6 25. 24 26. 6 27. 16 28. 5 29. 10 30. 2 31. 0 32. 20 33. 7 34. 31 35. 11 36. 30 37. 11 38. 4 39. 36 40. 0
41. 7 42. 0 43. 33 44. 31 45. 9 46. 9 47. 25 48. 24 49. 21 50. 22 51. 7 52. 21 53. 7 54. 6 55. 13 56. 15 57. 35 58. 4 59. 12 60. 1

Day 96
1. 24 2. 6 3. 15 4. 28 5. 1 6. 18 7. 13 8. 2 9. 0 10. 20 11. 2 12. 1 13. 14 14. 13 15. 22 16. 22 17. 21 18. 1 19. 0 20. 1
21. 8 22. 23 23. 7 24. 1 25. 4 26. 2 27. 7 28. 17 29. 23 30. 4 31. 18 32. 7 33. 9 34. 9 35. 5 36. 2 37. 5 38. 11 39. 15 40. 5
41. 15 42. 20 43. 5 44. 27 45. 12 46. 1 47. 18 48. 5 49. 14 50. 27 51. 26 52. 2 53. 27 54. 4 55. 4 56. 9 57. 5 58. 29 59. 7 60. 18

Day 97
1. 7 2. 4 3. 5 4. 29 5. 24 6. 16 7. 14 8. 1 9. 22 10. 22 11. 4 12. 0 13. 18 14. 8 15. 28 16. 14 17. 24 18. 16 19. 2 20. 26
21. 3 22. 1 23. 26 24. 4 25. 24 26. 18 27. 5 28. 0 29. 1 30. 30 31. 4 32. 2 33. 8 34. 9 35. 3 36. 3 37. 27 38. 34 39. 0 40. 17
41. 2 42. 30 43. 28 44. 5 45. 14 46. 11 47. 11 48. 2 49. 11 50. 2 51. 21 52. 0 53. 27 54. 36 55. 12 56. 11 57. 2 58. 4 59. 18 60. 7

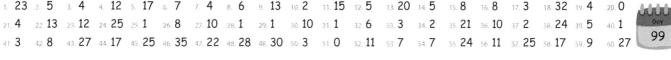

Day 98
1. 2 2. 2 3. 6 4. 0 5. 4 6. 26 7. 23 8. 16 9. 35 10. 28 11. 16 12. 3 13. 1 14. 26 15. 25 16. 31 17. 2 18. 2 19. 16 20. 31
21. 12 22. 8 23. 23 24. 14 25. 0 26. 12 27. 5 28. 16 29. 10 30. 23 31. 6 32. 5 33. 11 34. 1 35. 8 36. 22 37. 14 38. 18 39. 8 40. 29
41. 5 42. 11 43. 0 44. 12 45. 9 46. 7 47. 15 48. 2 48. 12 50. 18 51. 15 52. 22 53. 19 54. 13 55. 11 56. 11 57. 13 58. 2 59. 14 60. 1

Day 99
1. 23 2. 5 3. 4 4. 12 5. 17 6. 7 7. 4 8. 6 9. 13 10. 2 11. 15 12. 5 13. 20 14. 5 15. 8 16. 8 17. 3 18. 32 19. 4 20. 0
21. 4 22. 13 23. 12 24. 25 25. 1 26. 8 27. 10 28. 1 29. 1 30. 10 31. 1 32. 6 33. 3 34. 2 35. 21 36. 10 37. 2 38. 24 39. 5 40. 1
41. 3 42. 8 43. 27 44. 17 45. 25 46. 35 47. 22 48. 28 48. 30 50. 3 51. 0 52. 11 53. 7 54. 7 55. 24 56. 11 57. 25 58. 17 59. 9 60. 27

Day 100
1. 18 2. 5 3. 10 4. 6 5. 1 6. 22 7. 33 8. 27 9. 4 10. 36 11. 8 12. 0 13. 0 14. 0 15. 16 16. 16 17. 24 18. 4 19. 19 20. 7
21. 6 22. 11 23. 1 24. 2 25. 17 26. 30 27. 3 28. 22 29. 27 30. 22 31. 13 32. 18 33. 2 34. 23 35. 9 36. 8 37. 1 38. 1 39. 19 40. 7
41. 5 42. 33 43. 27 44. 17 45. 20 46. 4 47. 15 48. 13 48. 21 50. 7 51. 1 52. 11 53. 14 54. 11 55. 22 56. 28 57. 11 58. 25 59. 26 60. 6

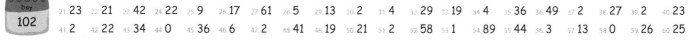

Day 101
1. 81 2. 95 3. 123 4. 83 5. 115 6. 61 7. 125 8. 114 9. 109 10. 135 11. 86 12. 57 13. 87 14. 79 15. 37 16. 54 17. 146 18. 138 19. 115 20. 92
21. 21 22. 164 23. 89 24. 146 25. 144 26. 85 27. 79 28. 110 29. 66 30. 101 31. 110 32. 89 33. 20 34. 60 35. 43 36. 73 37. 99 38. 51 39. 160 40. 182
41. 72 42. 80 43. 32 44. 81 45. 161 46. 91 47. 181 48. 99 48. 110 50. 52 51. 96 52. 71 53. 29 54. 109 55. 112 56. 83 57. 7 58. 92 59. 167 60. 117

Day 102
1. 74 2. 20 3. 30 4. 32 5. 55 6. 34 7. 72 8. 14 9. 31 10. 36 11. 2 12. 47 13. 41 14. 3 15. 1 16. 76 17. 59 18. 47 19. 12 20. 25
21. 23 22. 21 23. 42 24. 22 25. 9 26. 17 27. 61 28. 5 29. 13 30. 2 31. 4 32. 29 33. 19 34. 4 35. 36 36. 49 37. 2 38. 27 39. 2 40. 23
41. 2 42. 22 43. 34 44. 0 45. 36 46. 6 47. 2 48. 41 48. 19 50. 21 51. 2 52. 58 53. 1 54. 89 55. 44 56. 3 57. 13 58. 0 59. 26 60. 25

Day 103
1. 18 2. 108 3. 3 4. 76 5. 4 6. 173 7. 60 8. 83 9. 88 10. 116 11. 11 12. 109 13. 45 14. 0 15. 4 16. 18 17. 99 18. 124 19. 128 20. 14
21. 156 22. 13 23. 90 24. 17 25. 131 26. 1 27. 1 28. 54 29. 47 30. 3 31. 73 32. 107 33. 157 34. 57 35. 23 36. 147 37. 8 38. 106 39. 87 40. 87
41. 176 42. 43 43. 143 44. 41 45. 31 46. 38 47. 30 48. 8 48. 10 50. 26 51. 70 52. 67 53. 127 54. 1 55. 54 56. 35 57. 66 58. 90 59. 70 60. 174

Day 104
1. 35 2. 4 3. 123 4. 187 5. 108 6. 134 7. 7 8. 26 9. 89 10. 102 11. 8 12. 21 13. 185 14. 57 15. 53 16. 79 17. 12 18. 103 19. 157 20. 102
21. 24 22. 1 23. 40 24. 108 25. 61 26. 27 27. 44 28. 68 29. 71 30. 2 31. 0 32. 52 33. 10 34. 8 35. 49 36. 17 37. 0 38. 16 39. 41 40. 134
41. 104 42. 16 43. 23 44. 64 45. 127 46. 39 47. 109 48. 44 48. 86 50. 12 51. 139 52. 6 53. 14 54. 18 55. 145 56. 107 57. 0 58. 14 59. 3 60. 19

Made in United States
Orlando, FL
22 June 2022

19047638R00063